FLETCHERS
ON THE
FARM

FLETCHERS
ON THE
FARM

KELVIN & LIZ FLETCHER

HarperCollins*Publishers*

HarperCollins*Publishers*
1 London Bridge Street
London SE1 9GF

www.harpercollins.co.uk

HarperCollins*Publishers*
1st Floor, Watermarque Building, Ringsend Road
Dublin 4, Ireland

First published by HarperCollins*Publishers* 2022
This edition published 2023

1 3 5 7 9 10 8 6 4 2

A catalogue record of this book is
available from the British Library

ISBN 978-0-00-855852-9

Printed and bound in the UK using 100%
renewable electricity at CPI Group (UK) Ltd

This book is produced from independently certified FSC™ paper
to ensure responsible forest management.

For more information visit: www.harpercollins.co.uk/green

To you, Liz. My wife, my best friend, my childhood crush, the mother of our four beautiful children and the one person who makes all of this possible.

I love your laugh, I love your smile, I even love your bloody frown – but most of all, I just love you.

* * *

To K, I'll never be inspired by anyone as much as I'm inspired by you and to all the 'M's – you're my best achievement. I love you all.

CONTENTS

PROLOGUE
LATE NOVEMBER 2021

Peace at last. The camera crew has gone home and once again it's just us Fletchers on the farm: Kelvin, Liz, Marnie and Milo. Oh, and the bump. Despite saying loudly and often throughout the filming of our farming show that we would *not* be having another baby, Liz is twelve weeks pregnant with our third. It's nothing we can't handle. Not after everything the last couple of years have thrown at us. From *Strictly* to the pandemic to buying a 120-acre farm during lockdown, we've taken everything in our stride.

When we first moved to our little corner of the Peak District just eight short months ago, we had no idea what we were taking on. Now, outside in our fields forty sheep, three pigs and three alpacas are waiting for their breakfast. For the moment, though, there's time to relax as the early-morning light filters through the bedroom windows; time to take a deep breath and feel proud of all we've achieved so far before …

'Mummy! Daddy! The pigs are in the garden!'

… the curtain goes up on another crazy day.

In pyjamas and wellies, we chase the pigs back into their pen. The sheep want checking. The alpacas need feeding. The builders doing up the cottage we're hoping will be a holiday let need someone to look at a leak. Marnie and Milo need dressing. There are tears when Ginger the Cavapoochon gets hold of somebody's breakfast. The phone is ringing non-stop. Our Texel tup has broken out and got into a field of ewes. And now snow is gently starting to fall.

We drop the kids off at school on our way to the hospital in a flurry of sleet. Five minutes in the peaceful waiting room of the antenatal department where Liz is due to have her twelve-week scan feels like a day in a spa compared to what we've got done so far. Liz lays down on the table and the sonographer presses her wand to the bump.

'I've got some news for you,' she says.

After the morning we've had, we're not sure we're ready to hear it, but there's a hint of mischief in the sonographer's eyes as she turns from her monitor to tell us: 'You're having twins.'

CHAPTER ONE

A GREAT BIG ADVENTURE OR A TERRIBLE MISTAKE?

LIZ

So, how did we end up on a farm? It's a question I often asked myself in the early days. We're neither of us from farming families, Kelvin and I. We're both proper townies through and through. What's more, I'm allergic to just about every animal you can think of. Yet in the middle of the pandemic we moved from our comfortable modern house on a smart estate in Oldham near Manchester to a centuries-old farmhouse surrounded by nothing but fields as far as the eye can see. We weren't just moving house; we were changing our whole lives. And if our first day on the farm was anything to go by, we might just have been making a mistake.

It was snowing when we moved in. That light dusting of soft, white flakes might have made our new home look very pretty as we drove up the long driveway, but the ice underfoot in the farmyard made carrying boxes in from the removal van tricky, and it wasn't long before we noticed that it was no warmer inside the house than outside. The ground-source heat pump had packed up.

'Kelvin!' I wailed when I realised we had no heating and no hot water. 'What are we supposed to do?'

I ripped open one of the boxes containing the contents of my old wardrobe, hoping to find a cosy jumper. The party dresses and high-heeled shoes packed inside seemed like relics from another life, as did Kelvin's *Strictly* glitterball, which was perched on top of another box in the middle of the farmyard. It looked so out of place. Were we out of place too?

I couldn't find a jumper, so I wrapped a sparkly scarf around my neck. It would have to do. Kelvin didn't need to be told how important it was to get the heating fixed. Later that day, we'd be picking up our two children – Marnie and Milo – from their grandparents. I couldn't bathe them in cold water or put them to bed in a room where you could see your breath in the freezing air. What's more, Kelvin was flying to Budapest the following morning to shoot a drama with Sheridan Smith. It was an amazing opportunity for him, but the timing was terrible. He couldn't leave me in such a mess!

Looking out through our new kitchen window, I saw that Kelvin was deep in conversation with someone on a quad bike who was well bundled up against the weather. That quad-bike rider knew how to dress for the cold.

'That was our new neighbour, Gilly,' Kelvin said when he came inside.

'Great,' I said. 'Does she know where to find a heating engineer?'

'I promise it's all in hand.'

An engineer duly arrived and coaxed the heating back to life. As the warmth started to come back to the house, so did my sense of humour. Though not about the spiders. It seemed that a spider had already made its home in every drawer or cupboard I opened, sending me shrieking for the vacuum cleaner. The house had not been lived in for two years and I could only hope that the spiders were the scariest of the wildlife that had moved in since the last owner moved out.

*   *   *

Eventually, all the boxes were unloaded, we found a place for Kelvin's glitterball in the living room and I lined up my shoes in our thoroughly vacuumed bedroom.

'You can wear a different pair every day when you go out to feed the sheep,' Kelvin joked.

'Are we really getting sheep?' I asked. 'Can't we just have a couple more dogs?'

Ginger, our lockdown puppy, was having a very good day, investigating all the farmyard smells and rolling in the worst of them. She seemed very pleased with herself when she trotted into the kitchen covered in heaven only knew what.

'Ginger! You stink!'

'That's country life,' Kelvin told me.

Country life needed a generous squirt of Febreze. Would I ever get used to it? The smells, the mud, the muck?

Standing in front of the tumbledown cottage that came with our new place and thinking about the renovation work ahead, I felt a sudden wave of longing for our old house, all tidy and modern and painted in tranquil shades of white and grey, in our old neighbourhood, where we knew all our neighbours and had so many friends. What had we done?

Kelvin picked up on my uncertainty and gave me a hug.

'We've got a while before we need to pick the kids up,' he said. 'Let's go for a drive. Survey our new estate.'

'Estate?' That was a very grand title for the knackered old barns I was looking at.

We climbed into Kelvin's car, which had never been driven on anything rougher than a gravel drive, and set off on our first Fletchers' Farm tour. Prior to that day, I'd made only the briefest of visits to the farm – the pandemic had prevented anything more – and let Kelvin deal with all of the paperwork, while I got on with running family life. In truth, I'd concentrated more on how the house looked – whether it had a roof, for example! – so it didn't quite register how much land we'd bought or what it looked like. Acres and hectares were just words at the time.

To the front and side of the house was woodland – a tangle of bare winter branches.

'Our woods,' Kelvin told me. 'Those are oak trees.'

Behind the big metal-roofed shed was a snow-dusted paddock with a stable block.

'That's ours too.'

We drove past a field full of sheep, which dotted the frozen earth like fallen clouds.

'Our field,' said Kelvin. 'The sheep are Gilly's.'

And on and on until we reached a field at the very top of the hill in whose shadow our farmhouse was built. The Peak District stretched out in front of us, still and quiet in the soft winter light. As the setting sun streaked the sky pink and purple, Kelvin waved his arm theatrically.

'All of this,' he said. 'It's ours.'

The view was breathtaking, quite literally so. Not just because it was beautiful, even on that winter's afternoon. From up on that hill, the scale of what we had taken on was suddenly so completely and wholly overwhelming. Those fields, those trees … they were all ours now. Our responsibility. After a year of pandemic-related delays that made it seem as though it might never happen, we really did own a farm.

Kelvin put his arm around me.

'We're going to be so happy here,' he promised me. 'You, me, Marnie and Milo. Just think of the adventures we'll have. We can camp out in these fields in the summer. The kids will build tree houses. We'll get them a horse. We'll grow all our own vegetables. We'll get sheep and cows and pigs and make this place a serious business ...'

As Kelvin reminded me of all the reasons we'd decided to make the leap – for a better life, for financial independence, for freedom – I turned from the view to look at him and I could see in his happy, excited face the boy I fell in love with.

'It's going to be brilliant,' I agreed.

CHAPTER TWO
THE NEW BOY

KELVIN

All the best love stories begin with that moment when eyes meet across a crowded room, and ours is no exception. Except that our crowded-room moment didn't happen at a party or in a nightclub. It happened during assembly at Thorp Primary School, Royton, one morning in 1992.

'Girls and boys!' The headmistress, Mrs Hoggard, clapped her hands to get everyone's attention. She put her hand on my shoulder and guided me to stand just in front of her. 'We've got a new boy starting at Thorp Primary today. I'd like you all to welcome Kelvin Fletcher.'

I looked out at the sea of faces and felt dozens of eyes staring back at me. Who could I sit next to? Which of these kids would be my friends? I'd been a member of a drama club since I was six, but those Thorp Primary kids looked like a tougher audience than I'd encountered so far.

To my left, two girls sitting together – one blonde and one brunette – whispered and giggled behind their hands. Then the brunette turned back to look at me with a wide beaming smile that looked as though she meant it. Perhaps she would be kind.

'Kelvin,' said Mrs Hoggard. 'You can sit next to Chris Scholes.'

The boy in question shuffled up to make space for me.

'All right?' he said.

Safely in my new seat, I sneaked a glance at the girls who'd been whispering. I caught the blonde one's eye.

'That's Michelle Haining,' my new friend Chris told me. 'And next to her is Elizabeth Marsland.'

*　　*　　*

That day, my first day at Thorp Primary, was a big moment for me. I didn't want to be there. I'd been happy at my old school. I didn't want to move.

I'd spent the first eight years of my life in Derker, right in the middle of Oldham. Derker wasn't exactly the nicest area to live, but Mum and Dad were proud of the little house they'd bought on Derker's Evelyn Street and I loved living there.

My mum and dad, Karen and Warren, were a true love match. The story of the day they fell in love was one I'd heard hundreds of times growing up, but I never got tired of hearing it.

When they met, Dad was working as a diesel fitter by day and a taxi driver by night. Mum was working behind the bar at her parents' pub, The Boundary.. Whenever Dad went in for a drink – always half a shandy because he was driving; he was known as the Shandy King – he would try to get Mum chatting. But Dad wasn't the chattiest and Mum wasn't interested. She was going out with some lad who drove an Alfa Romeo. Dad didn't stand a chance against a lad with an Alfa Romeo. She might as well have been dating James Bond.

There was no way our dad could afford a flashy car back then. He had a four-year-old daughter, Keeley, working all hours to give her the best life he could. In the words of our mum, he had 'baggage'. Why would she want to get involved? But Dad never stopped hoping that one day Mum would see him differently and then fate threw him a chance to be a hero.

One night, while Dad was driving his taxi down Yorkshire Street, he found himself behind that legendary Alfa Romeo. It was driving erratically. The brake lights kept coming on. Then, all of a sudden, it slowed right down, the passenger door flew open and a woman tumbled out onto the road.

The Alfa drove off, but Dad jumped out of his car and ran to the passenger's aid. It was Karen from the pub – my mum. Dad helped her up, drove her to A&E and waited five hours for her to be treated. It wasn't just because she was his dream girl. Dad would have done the same for anyone. That's just the kind of man he is.

Eventually, Mum was patched up and Dad was able to see her. Though she was bruised and grazed and must have been feeling pretty sore, the only thing Mum really seemed worried about was that she was missing a shoe. It must have fallen off when she tumbled from the Alfa.

'That pair were my favourite!' she cried.

Without hesitation, Dad drove back to the spot where the incident had happened and searched high and low until he found that missing shoe in the gutter. He returned to the hospital like Prince Charming in search of Cinderella. The shoe fitted, of course, and the rest is history. A year later, they were married and Mum had become step-mum to Keeley. Then I came along, followed by my brother Dean and, a decade after that, our little brother Brayden.

*   *   *

We had a very happy childhood, my siblings and I. Mum and Dad worked hard to give us everything we needed and made sure we were never aware of how tough it was for them. They were always both working at more than one job and, at night, while we kids were sleeping, they'd be counting out coppers to get enough together to buy us treats or maybe go to the bingo and try to win a big prize.

Nobody in our neighbourhood had anything much, but the best things about living in Derker were things that money couldn't buy. The locals might have been rough around the edges, but they were hard workers, like my mum and dad, and there was a real sense of togetherness. Everyone looked out for each other, and we came together as a community to make some really good memories: like on 5 November when we'd build a big bonfire at the end of the road. All the adults would bring food to the party – parkin, tater ash, mushy peas – it seemed like a feast to us children.

When I was eight years old, Derker was the centre of my world, so when Mum and Dad told us we were moving to a new house in Royton, I actually cried. My brother Dean and I both protested against the move, but we couldn't persuade Mum and Dad to understand how big a mistake they were making. It's no exaggeration to say it was my first heartbreak, leaving the two-up two-down in Evelyn Street that I'd always called home. Though we were only going two miles down the road, it felt like we were leaving our Derker mates forever. How would I ever make such good friends again?

Fortunately for me, Michelle Haining and her friend Elizabeth Marsland weren't going to let me remain friendless for long. At lunchtime on my first day at Thorp Primary, they marched up to me in the playground.

'Are you going to be on our team for British Bulldog?' they asked.

It was less a question than an order, and from that moment forward I was part of their gang.

*   *   *

Life in Royton was very different from Derker. Our new house was like a palace compared to Evelyn Street. It was a three-bed semi with double glazing and a driveway of its own. The double glazing was a big deal for my parents. Back in Derker, I remember Mum and Dad getting out tracing paper and using it to stick fake lead batons on the windows to make it look like we had the real thing. At our new place, the lead was pressed between the two panes of glass, like it should be. We had fitted carpets, a leather suite in the living room and bunk beds for me and Dean.

The house at 10 Camberwell Way was posh and my new Royton friends were posh kids. They weren't streetwise like my old Derker mates. They definitely weren't starting fires for fun. I was used to kids who wanted to fight.

The most daring game we played on the Thorp Farm Estate in Royton was Knock-a-Door-Run. In Derker there were places we kids all knew not to go – places where the druggies hung out – but Royton felt really safe. In all my memories of those days, the sun is shining, the weather's warm. It's always summer. We'd get home from school, have our tea, then rush out to play. We'd play kerby with a football for hours, or just hang out in the local cemetery while the girls practised their roller-skating on the newly tarmacked paths between the graves. There was a stream nearby that we dared each other to jump over (Liz ended up falling in and getting soaked). We'd stay out until the street lights came on, which was our signal to go home. Though I'd been seriously upset about leaving Evelyn Street, I soon came to like Thorp Farm. I loved my new friends: Michelle, Matt Stirland and Scholesey. And Elizabeth Marsland, of course.

There was always something different about Liz. She stood out to me from the very beginning. I liked her warm smile and her curly bob. She teased me and made me laugh. She was my very first crush. When we played spin the bottle, I'd cross my fingers and hope we'd have to kiss.

For three years, our little gang of five was inseparable, but all good things come to an end. Liz went off to secondary school at North Chadderton. Michelle and Matt went elsewhere. Only me and Scholesey were left behind with one more year to go at Thorp Primary. Suddenly it was as though we were living on different planets. I would later go to North Chadderton too, but Liz didn't want to be seen with younger kids any more and so our gang drifted apart.

It stung a bit, knowing that Liz had moved on, but there were other things going on in my life by then.

Right before we left Derker, I'd joined the Oldham Theatre Workshop, a kids' drama club in the town centre. My big sister, Keeley, had been going there a while. When she took me along with her one Saturday morning, to get me out of Mum and Dad's hair, I had no idea how important that adventure would turn out to be.

CHAPTER THREE

NOT QUITE READY TO RHUMBLE ...

KELVIN

Though the Oldham Theatre Workshop drew kids from some of the poorest parts of town, it was a serious player in the national youth theatre scene. This was in large part due to the leadership of drama tutor David Johnson, who founded the workshop in 1968.

As I was to discover aged six, from the moment a child stepped through the door, David treated them like a professional actor and expected them to behave accordingly. He could be frightening – both to us kids and to our parents. He could look out over a sea of chattering children and silence them all with a couple of well-chosen words. The phrase 'You boy! You with the hair!' can still freeze me to the spot. But there was something about David that always made you want to do your best. He was big on discipline. If you wanted to act in one of his productions, you had to arrive on time and know your lines. You had to prove yourself.

David's workshops were transformational. Some of the kids would turn up wearing shoes with holes in – they were that poor – but on stage they always found a safe place where how much money their parents had or didn't have didn't matter. On stage we could be whoever we wanted to be. And with David pushing us on, everyone gave their all.

*   *   *

I was eight years old when David cast me as the lead in a play called *Charlie Is My Darling*. The workshop had been asked to send a group of kids to perform at the London Palladium, in a variety performance showcasing the best of youth theatre in the UK, and *Charlie Is My Darling* was going to be our play. It was a big deal for me in my first starring role.

Charlie Is My Darling was set in Victorian England. For my costume I wore ragged clothes and plastered soot over my face. If I do say so myself, I was a cute kid (despite the dodgy haircut Liz remembers from my first day at Thorp Primary – she claims my fringe was like a pair of curtains), and I was even cuter as a Victorian urchin.

The part of Charlie called for me to have a pet mouse. I could have used a toy mouse as a prop, but I decided I had to have the real thing for the sake of authenticity. Somehow I persuaded Mum that this was a good idea and she took me to the pet shop in search of a co-star. The pet shop owner immediately put us straight. A mouse would be too difficult to handle, not to mention difficult for the audience to see at a distance.

'What you want is a rat,' he said.

I'm not sure that what Mum wanted was a rat, but she was a good sport about it and we left the shop that afternoon with a rat I named Roland.

It was my first experience of looking after an animal. Over the next few weeks I bonded with Roland as I rehearsed my part over and over. That rat was a great co-star. He seemed to understand what he had to do. There was only one problem. During the performance, Roland had to stay quietly hidden in my pocket until the moment I got him out to show him to my character's love interest, played by a girl called Laura Dickinson. He was good at staying hidden. But every time I put my hand into my pocket to fetch him, I'd discover that he'd done a poo. It made it difficult to keep a straight face. It was worth the trouble, though. Everyone was talking about us afterwards.

*　　*　　*

There's no doubt that the Oldham Theatre Workshop punched above its weight, but compared to some of the fee-paying drama schools represented at the Palladium that night it was just a little youth club. One of the big players was the Sylvia Young Theatre School, whose former students dominated the professional acting world. After my performance as Charlie, I was invited by Sylvia Young to audition for a scholarship. I was amazed. Were they serious?

'You should give it a go,' said David.

It was going to be hard. I needed to be able to act, sing *and* dance. We'd just moved to Thorp and I'd joined the school choir. When I told Mrs Farmery, who led the choir, about the audition, she promised to help me prepare some songs. To give her an idea of the kind of help I needed, I sang an extract from ABBA's classic mini-musical *The Girl with the Golden Hair*.

I gave it everything, flinging myself into the song, acting out every line. As I hit the high notes in 'Thank You for the Music', Mrs Farmery did her best not to wince.

When I came to the end, she gently shook her head and said, 'Kelvin, you're going to need a lot of work.'

When it came to dancing, I had help much closer to home. My sister, Keeley, who'd just turned thirteen, was a great dancer. She spent most of her free time at the Marianne Jepson College of Dance & Drama, another

Oldham performing arts institution. Keeley agreed to choreograph a routine for me. We chose PJ & Duncan's 'Let's Get Ready to Rhumble' and she showed me some street moves that she thought would set me apart from the crowd. By the time Keeley had finished with me, I could certainly have cleared the dance floor at the school disco but for all the wrong reasons. Let's just say Ant and Dec had nothing to worry about.

When it came to acting, thanks to David Johnson, I already had that nailed. I rehearsed a monologue in which a young boy talks to his absent father. It felt emotional and authentic, which suited my acting style well. I couldn't wait to show the teachers at Sylvia Young what I could do.

On audition day, Mum and I travelled down to London on the train. I was fizzing with excitement as we got nearer to the capital. It's only looking back now that I understand the sacrifices my parents were making back then to support my performing ambitions. They went without all the time to pay for things like that train fare. They didn't go out, they didn't have Sunday lunch; they were always budgeting so that I could follow my dream.

*   *   *

I'd worked hard in the run-up to that day at Sylvia Young, but as soon as Mum and I walked into the hall I knew this audition was going to be a different kind of gravy. It didn't take long to see that I was the odd one out. I couldn't hear any other northern accents. The clothes I wore were different too. I'd arrived to find the other kids already warming up, wearing proper leotards and dancing shoes – the kind we couldn't afford. I was in my tracksuit and trainers. Mum tried to keep my confidence up, but I'm sure she was wondering, as I was, whether this was all a big mistake.

Since the other kids were practising their dance routines, I decided to practise mine too. After a few minutes, Mum gently took me to one side and told me, 'Best not rehearse your moves out here, Kelvin, love. Somebody might try to copy them.' In reality, I think she was trying to stop me from embarrassing myself. I was like a keen amateur football player turning up at a premiership club's training ground.

Though my dancing and singing weren't the best that day, I did get through to the last nine out of a field of hundreds. On the way back to Oldham, I felt as though I might have a chance, but a couple of weeks later I got a letter telling me that the scholarship had gone to someone else.

It was not my first rejection and it wouldn't be my last, but it was hard to have got so far only to be pipped at the post. I cheered myself up with the fact that the rejection letter noted that my acting had been 'nothing short of outstanding'. Mum also reminded me that I'd have had to move to London if I studied at Sylvia Young. I didn't feel ready for that.

'There will be other opportunities,' David Johnson promised me.

*   *   *

David's brand of tough love worked. We didn't know as children how well he was setting us up for life in the acting industry. What seemed like harsh lessons at eight years old put us one step ahead as young performers. When we went out into the real world, we felt we had the edge. David championed us. He pushed and pushed, encouraging us to find more with every performance.

'Energy!' he'd cry and we'd take it up another notch.

One look at the names of the students who started out with him tells you just how good David's methods were: Anne Kirkbride, Sarah Lancashire, Anna Friel, Antony Cotton, Lisa Riley and Suranne Jones. They all began their careers at the Oldham Theatre Workshop.

Though I had missed out on the Sylvia Young scholarship, it wasn't long before I was starting to get professional acting work. I got small parts in *Heartbeat* and *Coronation Street*. I'll never forget how exciting it was to be cast in a commercial for Persil. Mum was at home when the call came in and she rushed to school to tell me. When I saw her standing at the edge of the playing field where I was doing cross-country, waving her arms to get my attention, I thought I must have forgotten a dentist's appointment.

'You're going to South Africa!' Mum yelled.

I was stunned. I'd never been beyond Europe before. I got a bit of stick at school for it, but even though I was still only nine years old, it convinced me that acting was the life for me.

CHAPTER FOUR
A BUSY BEE

LIZ

I came from a family where school and education were really important. Dad was a lecturer in electronics. My mum, Mary, went to university as a mature student. She was in the middle of her law degree when she became pregnant with me. She worked hard to graduate while looking after three children and quickly rose to be a senior Crown prosecutor for the Crown Prosecution Service. My big brothers Daniel and Michael were both very academic (they went on to become a surgeon and a dentist). But me? I was always a performer.

As soon as I could talk, I liked to put on a show. Mum often tells the story of my first public appearance, playing a 'busy bee' in a show with the Marianne Jepson College of Dance & Drama, which was very well known in the area for teaching performers discipline and technique even from a very young age, none of which I ever followed! At rehearsals, I'd been given a place right at the back of the chorus. Well, I wasn't having that. As soon as the curtains opened, I barged straight to the front of the stage.

'Ta-daa!'

No matter what I was supposed to be doing, I always found an opportunity to get into a role. Sometimes, when Mum was unable to get childcare during the school holidays, she would take me into work with her. I'd answer phone calls, putting on my best telephone voice, pretending to be a secretary three times my age.

My passion for performing only grew as I got older. I loved singing and persuaded my parents to let me have lessons. I did ballet classes and acting classes. I learned how to play the violin and piano. All that tuition must have cost a fortune. We had one car, which my mum would use, so Dad would take me to school on his bicycle, sitting on a cushion on his crossbar. I thought it was the funnest way to travel.

My best friend Michelle Haining was mad about all the same things. In a primary school performance of *Oliver!* We played Nancy and Bet. We made a good double act. We stayed friends beyond primary school, and during the many school holidays, Michelle and I would spend our time over at her house, watching Sky. I remember one year we were obsessed with Britney Spears in the video for

'Baby One More Time' and spent many hours trying to copy her dance routines. We made up routines of our own too – strutting and posing like dancers twice our age in our mums' high heels – and filmed the best ones on Michelle's dad's camcorder. If we weren't pretending to be Britney, we were pretending to be backing singers for Peter Andre, another big favourite of ours, or we were imagining ourselves as the stars of our own girl band. We'd cover ourselves in make-up and put on shows, charging our friends and families twenty pence to see us dance and sing. I'm sure that having listened to us practising our Britney tunes day in, day out, some of them would have paid forty pence *not* to have to be in the audience.

*   *   *

Like Kelvin, I had my first taste of professional theatre before I left primary school. Our school choir was invited to audition to take part in a run of *Joseph and the Amazing Technicolor Dreamcoat* at Manchester's Palace Theatre. Michelle and I were over the moon when we were picked to be in the show.

It was so exciting. We couldn't wait to get on the coach that took us to the theatre, where we rehearsed alongside cast members. We didn't get much time with the big stars, but it was thrilling to be on stage with professionals and see how a real theatre worked.

My family were really supportive of the whole experience, making sure that for every performance there was always someone I knew in the audience, whether that was a family member or a neighbour. They encouraged everyone they knew to come along and by the time the show's Manchester run ended, I was addicted to being on stage.

I kept taking acting lessons and around the age of thirteen I joined a local drama club. We put on *The Wizard of Oz* at the Queen Elizabeth Hall in Oldham and I played Dorothy. Towards the end of the run, David Johnson came to see the show.

After he saw me on stage, David wrote to the leaders of the drama club, telling them he wanted me to join his acting classes. By this time, David had left the Oldham Theatre Workshop and set up on his own at the Ragged School in central Manchester. It was a prestigious invitation that I was delighted to accept.

It was everything I wanted – to be asked to join such a renowned drama school – but in the lead-up to my first Saturday-morning class, I was so anxious I could have thrown up. Walking in, I wanted to melt into the floor. The other kids seemed so confident. Many of them were three or four years older than I was. They towered over me. I didn't know how to compete.

When David asked us to find another person to partner up with, I stood there rigid with nerves. No one was rushing to pair up with me. Was I going to be stuck on my own? You can imagine my relief when a familiar face from school appeared from the crowd of strangers.

'Do you want to be my partner, Liz?' Kelvin Fletcher asked, smiling his big, toothy smile – the same smile that had charmed me when I first saw him standing in front of the class at Thorp Primary.

'I suppose so,' I said, trying to be much cooler than I felt.

Our task that day was to improvise a scenario based on the theme of 'intimidation'. I almost burst out laughing at the idea. Intimidation? Kelvin and I were the smallest kids in the class. Between us, we couldn't have intimidated a fly. Still, we both wanted to impress David, so we came up with a scene based around a job interview. I would play the interviewer, Kelvin the interviewee. So far, so straight-forward. Perhaps I should intimidate him with difficult interview questions? No. Kelvin had a better idea. Halfway through our interview, he would pretend to pull out a knife.

I thought he was crazy but his enthusiasm for the idea carried me along.

'It's going to be really good,' he said.

He'd found a plastic knife somewhere that he used as a prop. When he dragged the blade down a leather chair we were using in our scene, he made a very convincing nutter.

I'd known what was going to happen but even I was shocked. There was a moment of horrified silence before I broke the tension in the room by pretending to tidy up some papers on my desk and saying, 'Err, well, someone will be in touch then ...!'

David was pleased with how Kelvin and I worked together, but it was to be a one-off. After that first day I didn't partner with Kelvin again, preferring to pair up with other kids who seemed much more interesting than my familiar old Thorp Primary friend, but looking back, I can see in that improvised scene how well we worked as a team even then.

*　　*　　*

Every Saturday, when my acting lesson finished, Mum picked me up to take me into the city centre. We had a routine that we stuck to every week. We'd pick up *Hello!* magazine from a newsagent, then go to the café at Kendals Department Store for a cup of tea and some teacakes while we read the celebrity interviews. I loved reading *Hello!* for the amazing dresses, looking for the perfect one I'd wear on the red carpet.

'And who are you wearing, Liz?'

'Well, obviously it's Chanel ...'

After that Mum and I would go shopping. It was our mother-and-daughter time.

On Saturday, 15 June 1996, Mum collected me from the Ragged School as usual and we headed into town. Back then, Manchester was a city of glass. Loads of the old derelict buildings around the city centre had been renovated in a really modern style with great big windows. Tramlines had been put in down the main streets. Everything was shiny and new.

I remember looking up at Key 103, the radio station, as we passed it in Mum's car – that building was all windows – when we suddenly heard and felt an explosion and watched, as if in slow motion, a huge glass pane fall from Key 103 onto a young woman in a wheelchair and her companion below.

It was terrifying. Unaware of what had been unfolding in the city centre, Mum thought she must have accidentally driven over a tramline and caused a short in the electricity circuit that somehow shook the windowpane loose. As we sat there, stunned, sirens started wailing as ambulances and police cars rushed to attend to the people caught up in the chaos, including that poor woman in her wheelchair.

'What's happening, Mum?' I asked, my voice shaking. The air was so full of dust that it was hard to see. 'Mum? I'm scared, Mum! What's going on?'

'It'll be OK,' Mum said, but her tone was unconvincing.

Still not knowing what had happened but realising it was nothing to do with the trams, Mum quickly turned the car around and we joined the rush of people leaving town. We were almost home before a news bulletin on the car radio revealed what was going on – the IRA had detonated a bomb in Corporation Street, right in the city centre.

When we got to our house, we found Dad frantic with worry. He rushed out onto the driveway as soon as he saw us turn in. He'd been listening to the news, unable to do anything but pray we hadn't been nearby when the bomb went off. We hugged each other very tightly that day.

CHAPTER FIVE

THIS IS OUR ANDY HOPWOOD

KELVIN

I was also at the Ragged School that Saturday in 1996. My class was right after Liz's. While David Johnson was putting us through our paces, a whisper went round that there'd been an IRA bomb threat to Manchester city centre and people were being evacuated. The IRA had targeted Manchester before, so we knew that it could happen. But before we could get confirmation of the rumour, we heard an almighty *whoomphf* and the windows of the old building where we rehearsed were blown in, covering the room with broken glass.

All hell broke loose. There was massive panic, with lots of screaming and crying. Only David's reassuring voice cut through the mayhem. Never had the discipline he demanded of his students been more valuable than it was right then.

'Settle down!' he demanded.

The adults checked us kids over and made sure we were all safe. None of us was hurt, though of course we must all have been in shock. And everyone was worried as to what might happen next. This was in the days before mobile phones, so we crowded around the one phone in the building, taking it in turns to ring our parents and let them know we were all right. I got through to Dad, who was about to come and find me. I wasn't crying but I was pumped full of adrenalin at the close call. The Ragged School was only five minutes from Corporation Street where the bomb had gone off.

The amazing thing about that day is that, despite the size of the explosion – it was a 1,500kg lorry bomb – and the enormous structural damage it caused, no one lost their life. Unfortunately, 200 people were injured, but no one was killed. That's thanks to the emergency services who managed to evacuate 75,000 people in the time between the IRA giving a coded warning and the bomb going off. It was a miracle, really, that so many lives were saved.

The emergency services did a brilliant job and the city stood firm. After the bombing, the centre of Manchester was totally rebuilt. They kept a single post box that had survived the bombing as a memorial. Whenever I walk past it, I remember that day and how David helped us get through it.

*   *   *

Of course the Manchester bombing was all anyone could talk about at school on Monday morning – everyone had their own story – but it wasn't long before life started to get back to normal. The windows at the Ragged School were replaced in a matter of days and it was back to business as usual.

My Saturdays always started with a session at the Ragged School. I'd do a two-hour class in the morning, then Dad would pick me up and take me with him to his work at Salford Van Hire. I'd spend the afternoon helping him out there. When I say 'helping him out', I mean Dad would give me the key to his office and I'd nick all the tokens for the coffee machine, so I could get unlimited hot chocolate. Dad was OK with that. He was less happy when I started pinching all the metal dust caps off the hire vans' tyres. There was a craze at school for collecting dust caps. I don't know why. We didn't do anything with them except gather as many as we could. The metal van ones were considered much more valuable than the plastic ones you got on ordinary car tyres.

*   *   *

Whereas the Oldham Theatre Workshop had been very 'Oldham', drawing the kids who went there from the local area, the Ragged School brought together kids from all over Greater Manchester. If you see a famous actor from the North on TV, if they weren't at the Oldham Theatre Workshop, they were at the Ragged School, which boasts Ryan Thomas, Tina O'Brien, Alan Halsall and Brooke Vincent among its former pupils. It was a hotbed of Mancunian talent. Ragged School students were always getting TV and theatre roles.

Shortly before the Manchester bombing, I'd got my own big break. I'd first met Sue Jackson, the Granada TV casting director, when I was eight. She was casting the part of Jack Duckworth's grandson, Jamie Armstrong, for *Coronation Street*. I got down to the final two, but the part ultimately went to Joe Gilgun, who'd been at the Oldham Theatre Workshop with me. It was a big disappointment, missing out on that part, so when Sue asked to see me for a part in *Emmerdale*, I reined in my expectations.

I turned up to the audition in Leeds wearing a fleece, with my collar up. I must have looked a right tearaway. Later, Sue would tell me that the moment she saw me wearing my fleece over my nose, she said to herself, 'This is our Andy Hopwood.'

* * *

Two long weeks after that audition, I heard that the role of Andy Hopwood was mine. When the call came, the whole family gathered round to hear the news. There was screaming and crying – happy crying, of course. It felt like my getting the part was a triumph for all of us: for me, for Mum and Dad, for my grandparents and my brother Dean and my big sister, Keeley. I knew I wouldn't have got so far without their support. They were 100 per cent behind me. Nan had already started a collection of tapes of every TV appearance I'd ever made, whether it was in a show or a split-second part in an ad. Now she'd have *Emmerdale* to add to her collection.

To begin with, I was going to be playing Andy Hopwood in just three episodes. His storyline went like this: Andy had a difficult upbringing. His mother, Trisha, died when he was small and his father, Billy, was in Strangeways Prison, leaving Andy to be brought up by his grandmother. In his first episode, Andy joined a group of kids being taken by social workers on a camping trip at Jack Sugden's farm. Once there, he found an instant connection with the countryside and also with Jack Sugden's son Robert. Robert took Andy off on an adventure that almost ended in disaster when Andy had an asthma attack. Jack Sugden stepped in to save his life.

Though in many ways, I was nothing like Andy Hopwood, in one way he and I were similar. We were both urban kids through and through. Though you didn't have to go far from Oldham to find glorious countryside – it lies on the edge of the Peak District National Park – I hadn't spent much time out there in the hills and I definitely didn't know anything about the farming life. Would I feel that same instant connection with the wild landscape that had been written into the script for Andy?

*　　*　　*

As my first day on the *Emmerdale* set approached, I was nervous and excited. One of the first people I met on set was Clive Hornby, who played Jack Sugden. Clive was a soap legend. Even people who didn't watch *Emmerdale* would recognise the character of Jack Sugden in his iconic flat cap and wax jacket. Clive was properly famous. He was also a man mountain. He towered over me as he shook my hand.

'Welcome to *Emmerdale*, Kelvin.'

Though he was known for playing a Yorkshireman, Clive was actually a Liverpudlian. Before he became an actor, he'd been in a band called the Dennisons who had shared a stage with the Beatles at Liverpool's Cavern Club. When the band split up, Clive worked backstage at the Liverpool Playhouse, which was where he was inspired to try his hand at acting. He was a 'proper' actor, trained at LAMDA (the London Academy of Music and Dramatic Art).

I think Dad was almost as nervous as I was when we met Clive for the first time, but Clive couldn't have been nicer. He had a son of his own, a little younger than I was, which must have helped him know how to put me at ease. But as soon as we started acting, Clive treated me like a peer, like he trusted me to know what to do and expected me to be professional. I pulled together everything I'd learned from David Johnson's workshops and gave it my all.

Playing those early scenes felt like real showbusiness. *This is all I want to do*, I thought, as Dad drove us back home afterwards. Like Andy Hopwood, I felt like I'd found my place among those rolling green Emmerdale fields.

I made the most of those three episodes, determined to learn everything I could from the experience. I thought that would be it for my time on *Emmerdale* – Andy Hopwood had come and gone – but a couple of months later, we heard rumours that the producers were thinking of using the character again. There was talk of a storyline in which Andy's grandmother died of a heart attack, leaving him without a guardian, and the Sugden family stepped in to foster him on their farm. I was really hopeful that the rumour would come true, but it was a delicate moment. I felt as though even daring to speak the words out loud might ruin everything.

But the rumours did come true, and it wasn't long before I was back at the *Emmerdale* studios again. And that was the start of a twenty-year-long adventure.

CHAPTER SIX

JUST SAY 'GEORGE'

LIZ

Kelvin was having much better luck than I was in getting his career off the ground, but I was still determined to give it a go. Since the age of twelve, I'd had a subscription to *The Stage* – the weekly trade magazine for people in the entertainment industry – and avidly scanned the auditions listed in the back for something I could try for. It was a great way to find auditions, though I wasn't the only one looking and would sometimes turn up to an open casting to discover that 200 girls my age and height had had exactly the same idea. Not only that: some of them had serious stage parents who were determined that their little darlings would succeed. I would turn up on my own as my mum and dad were always working. I didn't really have a clue what I was supposed to be doing.

By this time, Kelvin was already having time off school to record *Emmerdale*. He was starting at the very top, on a show that had millions of viewers. Of course, being on *Emmerdale* made him something of a celebrity at school, but whenever I talked to him he was just the same old Kelvin I'd known for years.

Kelvin and I did star in one show together while we were still at secondary school. It was *Bugsy Malone*. Kelvin played the eponymous Bugsy. He wasn't actually the drama teacher's first choice for the part, but when the original Bugsy dropped out partway through rehearsals, the drama teacher knew that if anyone was going to be able to get up to speed quickly and save the show, it was Kelvin Fletcher.

Since the moment I heard that *Bugsy* was going to be our school play, I'd been determined to win the part of Blousey, the female lead and Bugsy's love interest. The part was made for me, so I was seriously disappointed when Blousey was given to Joanna Humphreys. With Kelvin playing Bugsy, the casting seemed plain ridiculous. Apart from anything else, Joanna was almost six feet tall to Kelvin's four feet nothing. They made a very odd-looking couple. But the drama teachers had made up their minds. I was given the much smaller role of Lena Marelli, the part played by Bonnie Langford in the film.

To make matters worse, Kelvin and Joanna started going out off-stage. Though I didn't think it mattered to me that much at the time, it was annoying to be around them making eyes at each other. Still, it spurred me on to make sure that the one full minute I had centre stage as Lena was the best minute of the entire musical. And to this day, Kelvin and I joke about Bugsy and Blousey.

'Are you over that girl yet?' I say.

*   *   *

That same year, an audition for my dream role appeared in the listings in the back of *The Stage*. It was for the part of Anne in a national tour of *The Famous Five* musical. I knew I had to go for it.

When my parents dropped me off at the first round, there were hundreds of other girls there with the same dream. Nevertheless, I stayed confident, determined that this part was mine.

The only problem was that the director wanted Anne to have an 'RP' accent. That's RP for 'received pronunciation', the kind of accent you associate with the Royal Family and old-fashioned newsreaders. My accent was definitely not RP. It was purest Oldham. It's an accent that's regularly voted one of the friendliest in the UK, but it wasn't right for the part of one of Enid Blyton's adventurous young toffs.

I would not let the small issue of my accent get in the way. I gave the audition my best and went home pretty sure I'd get a call-back. I remember the day the call came.

I was sitting in class at school when I got a really strong feeling that *The Famous Five* director was going to ring. I've always had this sort of intuition when something big is going to happen. We didn't have an answer machine in those days and nobody in our family had a mobile, so the moment I got home I dialled 1471 to see if anyone had phoned while the house was empty. The last call was from a London number. With my heart in my mouth, I called back. My intuition was right. It was *The Famous Five*'s director, Dan Crawford.

'We want you to come to London for another audition,' he told me.

I couldn't wait for Mum and Dad to get home from work so I could tell them what had happened. The auditions, which were to be held at the King's Head Theatre in Islington, were just a couple of days away. Dad's mum – Grandma – volunteered to take me down on the train.

The second round was much more nerve-wracking than the first. One of Enid Blyton's daughters was there with Dan and his team, keen to make sure that her mother's most famous characters were brought to life in a way that fitted her mother's vision. More than ever, I needed to get it right.

During the course of the day, I was asked to act with various other people in a variety of scenes from the play. I was sure I was capturing the spirit of Enid Blyton's Anne in my performance, but the problem remained my accent. There was one word in particular that brought my Oldham vowels to the fore. It was 'George'. Unfortunately, George the tomboy was a character whose name I would have to say. A lot.

'Just say *George*,' Dan kept saying, pronouncing the word in a way that was totally alien to me.

'George. George. George. George?'

I could hear what Dan was saying but there seemed to be no way to get my mouth to make the same sounds. I felt like Eliza in *My Fair Lady*, learning how to 'speak proper' for Henry Higgins.

'*Ge-or-ge*,' Dan tried again.

Once more, I tried and failed.

I couldn't do it. I felt my chance of playing Anne slipping away.

Finally, in exasperation, Dan held up a coin – a special coin I'd never seen before with King George's face on it. 'This could be yours, Liz. Just say George properly.'

I remember being so moved by this gesture. In some way I felt the coin represented his belief in me, that I could do it, and I really wanted to make him proud of me. I must have said George a thousand times that afternoon, but finally I did it.

'You've got the part,' said Dan.

*   *   *

To this day I still remember the feeling of pride that I'd nailed it. Over the next few days, I got ready to join the tour. The rehearsals were to take place over four weeks in London, where I would have to stay with a host family. It was exciting and terrifying all at once. It was to be the first time I'd ever really spent any time away from my parents. Despite my big ambitions, I was actually something of a homebody. If I ever stayed at a friend's house for a sleepover, I'd inevitably end up calling my parents at three in the morning, begging them to pick me up so I could come home and sleep in my own room because I was so homesick. How on earth was I going to get through four weeks staying with total strangers?

Dan's team assured my parents that the host families would look after us as though we were their own children. I was placed with a woman who had the strongest Cockney accent I'd ever heard. When her kids were in trouble, the whole street knew about it. Meanwhile, Mum and Dad decided it was time I had my first mobile phone, so I could call home whenever I wanted. That should make me feel less alone. The only problem was that the phone was huge, as big as a brick with a SIM card the size of a credit card, and I was embarrassed to get it out of my bag – the result being that Mum and Dad still had to call the house where I was staying to make sure that I was OK.

Despite my anxieties about being away from home, I soon settled in. Rehearsals kept me busy during the day and the other cast members became good friends. One of them was Lyndon Ogbourne, who would end up playing Nathan Wylde on *Emmerdale*. Then there was Olivia Hallinan, who went on to appear in *Holby City*, *Doctors* and *Sugar Rush*. Though life has taken us in different directions over the years, we three have remained in touch.

My teachers back in Oldham sent schoolwork via my parents, in an attempt to make sure I kept up with my studies, but I was away for twelve months and I did the bare minimum. I was much more excited about my first professional role in musical theatre than in keeping up with maths and science.

I didn't miss school at all. I'd always been smaller than most of my peers and found the rough and tumble of the playground a bit scary. I hated the school politics too. I wasn't shy, but neither was I in with the 'in crowd'. Being part of their gang involved showing you could be rebellious, but I didn't want to go drinking in the park at weekends.

The most rebellious thing I'd done so far was trying a bit of joyriding. Well, it wasn't exactly joyriding, since we didn't really go anywhere. On Christmas Eve, my friend Sally Johnson and I were sitting in her mum's car, waiting for her mum to do some errands. Getting bored, we decided we'd try to drive. We got the car into gear, pressed the accelerator and bunny-hopped straight into the back of the white van parked in front of us. We'd hardly moved an inch but we'd crumpled the van's bumper. The big burly guys who got out of the van were ready to kick off until they saw two little girls behind the wheel. They were very kind about it, but you can imagine how pleased Sally's mum and my parents were. That was my joyriding career done and dusted.

In her role at the Crown Prosecution Service, Mum was starting to see some of my schoolmates coming through the criminal justice system for petty crime. I can't imagine what she'd have done if I'd followed that path. I'm sure my parents thought I was safer on tour!

*   *   *

Once rehearsals were over and *The Famous Five* was on the road, my family and friends saw the show whenever they could. My best friends back in Oldham were really supportive. When the tour got to the Manchester Palace, practically my whole class came to watch. At the end of the show I was leaving by the stage door and the whole class jumped on me and piled around me in one big excited group hug. Michelle said she loved seeing my face in the show's programme. Dad would get very emotional whenever he was in the audience watching me take a bow.

I was so happy to be making Mum and Dad proud. My brothers were always getting A-stars in their exams. They were off to university. This was my own unique way to succeed. It was my first job representing a show as the lead. We even got to talk to Richard and Judy on *This Morning* to promote it – a big deal for a thirteen-year-old. I'll never forget opening the programme to see long lists of work the other kids had done, only to find my entry and a small paragraph under my name. Still, even though I was a little embarrassed at the time, deep down I was proud that little young me was standing among all these trained kids.

A year later the tour was over and I returned to Oldham. Though I'd been speaking with that hard-won RP accent night after night for the best part of twelve months, as soon as I was back home and at school, my old accent was back and you'd never have known I'd left Oldham at all.

CHAPTER SEVEN

AN AWARD-WINNING ACTOR

KELVIN

Being on *Emmerdale* was a dream job for me. It was a really warm, lovely place to work. Shortly after I made my first appearance as Andy Hopwood, the show had moved to a new studio in the centre of Leeds. It was a purpose-built studio, like the set Granada already had for *Coronation Street*. Meanwhile, a mock-up of the Emmerdale village was put up on the Harewood Estate. It was modelled on Esholt in West Yorkshire, the real village where *Emmerdale*'s exterior shots had been filmed for almost twenty-two years. The replica houses were full size, but inside were not the interiors *Emmerdale* fans might have expected; there was the make-up department, the equipment stores, toilets and offices for the production team and crew.

I probably spent 60 per cent of my time in the studio, filming interior scenes, but I definitely preferred location days at the 'village'. The drive to the village was beautiful. We'd park up at the base facilities, then walk down that famous street past the pretty sandstone cottages. The best moments were early on a summer's morning, when the sun was rising and the birds were singing. It was so picturesque at that time of day and felt so real. I couldn't think of a lovelier walk to work. Though *Emmerdale* was always full of drama, there was something calming about being in the countryside and spending time with the animals. I loved talking to the farming specialists who worked as advisers on the show, making sure that when any of us actors had to interact with a cow, for example, we looked like we knew what we were doing.

It was a great experience playing a farmer on set. Every time I did a scene with the Sugdens' cows, I came away wondering whether it was such good fun to be a farmer for real. Did they live the life I imagined?

There was always a sense of camaraderie and humour on set and practical jokes were common. That suited me. I was always a joker. When I was ten, I got a part on a Granada TV show called *Three Seven Eleven*, set in a fictional primary school. Filming took place in Warrington. While hanging about between takes, I discovered that if I scuffed my feet along the synthetic carpet in the dressing room, I could build up enough static to give the next person I touched an electric shock. Soon, all the kids on the show were doing it to each other and the grown-ups. It drove our chaperone mad.

I hadn't changed much by the time I got to *Emmerdale*. Though when it mattered I knew how to be professional, I also knew how to have fun. Woolpack scenes in particular were a great opportunity to have a laugh. Lots of our jokes revolved around breaking wind – it was always entertaining to force a loud trump when going for a serious take.

Some jokes took a bit of preparation. Once I made sure I got to the canteen for tea break before everyone else and put holes in the bottoms of all the polystyrene coffee cups with a tea stirrer. Just tiny holes. Not so big that they would mean anyone filling a cup with hot liquid got a scalding, just big enough that everyone who got one of the pierced cups ended up wondering where the annoying drip was coming from.

I enjoyed being the butt of jokes too. If you give it, you've got to be able to take it. Clive Hornby was a devil with the practical jokes. When I had to do scenes with him and he was off camera, he'd keep pulling faces until I started corpsing (acting slang for cracking up), knowing I'd laugh my head off and be the one who got into trouble.

Even the *Emmerdale* animals got in on the act. Once, Clive and I were filming a serious scene with one of the Sugdens' cows in the background. The cow had its backside towards the camera as Clive and I talked. We were in the middle of what seemed like a perfect take when the cow started dropping a cowpat. We were used to that and maintained our focus, just as any real farmer would. But halfway through pooing, the cow suddenly coughed, which triggered a fart that sent cow shit flying – over me, over Clive and over the cameraman. We were all covered.

'Never mind,' said Clive as he wiped cow muck from his cap. 'We can send the tape in to *It'll Be Alright on the Night* and make £300.'

I made a note to make sure I was standing at the head end of that cow next time I shared a scene with her.

We had a lot of laughs, Clive and I, but if I needed picking up on my behaviour, Clive was willing to do that too. On one occasion, I was late for work. I was travelling in from Manchester and lost track of time. By the time I arrived, the crew had been waiting for me for more than half an hour. It was a serious issue, wasting so many people's time. Between takes, Clive took me to one side and said, 'Kelvin, you don't do that. It's page one.' His disappointment in me was enough to make sure I was never late again without a really good reason. I hated the thought that I had let him down. It mattered to me what Clive thought of me, possibly more than he knew.

The way Clive handled my mistakes as a young actor taught me a great deal about how to handle problems that cropped up on set. You address it with an honest chat and move on, just as in family life. Clive's kindness and understanding made my admiration for him grow ten-fold. He felt like family on screen and off.

*   *   *

I got some big storylines in my early years on *Emmerdale*. In one, shortly after Andy was fostered by the Sugdens, his father, Billy, came back to the village.

Billy Hopwood was a wrong'un. The storyline around his return had some fantastic scenes, including one in which Andy pleaded with Billy not to mess things up for him with the Sugdens. 'I've got a family now,' Andy told him.

At this stage I was twelve years old, with some seriously big emotions to act. Should Andy help his real father rob the Emmerdale post office or remain loyal to the Sugdens, who'd become the family he'd always wanted and needed? It was a life-changing moral dilemma. Andy's decision led to another scene with a heated row, which ended with Billy slapping him. Andy chose the Sugdens.

But Andy's loyalty to the Sugdens wasn't always a force for good. The year I turned fifteen, I was given a storyline that would stretch me to the max. With Emmerdale Farm facing financial ruin, Andy set a barn alight, thinking the Sugdens could claim on the insurance. It was meant to be a victimless crime, but Andy was unaware that his adoptive mother Sarah was inside the barn when it caught fire. Sarah was killed.

I played every scene from the heart, and it got me noticed. Thanks to the fight scenes with Billy, I was nominated for best actor at the inaugural British Soap Awards.

When the *Emmerdale* producers told me I'd been shortlisted, I was speechless. There were no special categories for child actors back then, so I was up against adults I regarded as idols: Dean Sullivan, who played Jimmy Corkhill on *Brookside*, David Neilson who played café owner Roy Cropper on *Coronation Street* and Ross Kemp, the *EastEnders* legend. I was just a kid. I couldn't believe I was even allowed to be on the same list. As the icing on the cake, I was also nominated for that year's 'best dramatic performance'.

The Soap Awards ceremony took place at the BBC Television Centre in London. I took Mum and Dad along as my guests, pleased to be able to acknowledge how much my success was in no small part due to their dedication and support. When the thought of acting for a living seemed like a fantasy, they'd cheered me on. They'd made a lot of sacrifices so I could get to where I was, and I wanted them to know that I recognised how hard they'd worked for me.

The dress code for the ceremony was black tie, so Dad and I hired suits from a shop in Manchester while Mum bought herself an evening dress. We were put up for the night at the Grosvenor House hotel on Park Lane. It was the first time we'd stayed in such a fancy place. It felt like a dream.

On the day of the ceremony, I was excited but nervous. I was glad to have Mum and Dad there beside me, but at the same time – like any teenager – I was a little bit embarrassed by them too. My parents were both star-struck and kept pointing out the people they knew from the telly.

'Look, Kelvin, there's Dot Cotton!' said Dad.

'There's Barbara Windsor!' Mum squealed.

'Stop pointing and staring,' I hissed at them. I already knew how much I hated it when people pointed and stared at me when I was out and about. But Mum and Dad kept forgetting that we weren't sitting at home in our lounge. They were having a great time, seeing all these people we'd watched on television for years and years.

It was a strange moment for me. On the one hand, I was at that ceremony as a nominee and had every right to be there alongside those big-name stars. On the other hand, like Mum and Dad, I couldn't help but be a bit overawed by the fact that we were surrounded by people I'd only previously seen on the small screen – like the *EastEnders* and *Coronation Street* stars. Mum and Dad were right. It *was* pretty amazing to be seeing them in real life.

It was a great night. Richard Madeley and Judy Finnigan compered. I didn't win the best actor award – that went to Ross Kemp as Grant Mitchell – but I did win the award for best dramatic performance, beating *Corrie*'s Georgia Taylor (Toyah Battersby), *Hollyoaks*' Kerrie Taylor (Lucy Benson) and *EastEnders*' biggest star, Barbara Windsor. It was a big surprise. As I walked up onto the stage, I was overwhelmed with happiness. Lifting that trophy felt every bit as good as winning an Oscar to me. It was not just for me, but for the team at *Emmerdale* and my family, especially my mum and dad.

*   *   *

One of the strange things about acting for a living is that you can spend all day in really exciting and sometimes glamorous surroundings, but as soon as you get home life goes on as normal. For me, that meant that while I might have been an award-winning actor, I still had to do my homework.

When I started on *Emmerdale*, I was given the option of having tutors on set, but it didn't work out. Instead, my parents made an agreement with school that I would keep attending lessons whenever I could and catch up on what I had missed at the weekends. I still had to have a responsible adult by my side at all times on set, so my Grandad David – Mum's dad – became my chaperone. I was really close to Grandad David. During the school holidays, we'd pretty much move in with him and Nanny June for the whole six weeks. They'd sold their pub and moved back into the bakery business, which was where they'd started out. They had a shop called Wreghitt's in Hurst Cross. It was a real family affair, with everyone helping out when they could.

Grandad David would pick me up at 6.30 each morning, with a butty to eat in the car. I'd be in make-up by 7.30, then go straight onto the set to do my scenes. By 1 p.m. in the afternoon I'd be back at school. The other kids in my class were pretty good about letting me borrow their books to catch up on the lessons I didn't attend. Lauren Fitten was the best. She had the neatest handwriting. Thank you, Lauren!

Academically, I was doing all right. I had a good imagination, which helped me with my English. I also had a good work ethic. Sometimes that meant I would be up until two in the morning doing work. It did mean I missed out on some of the social life my schoolfriends were getting into as we got into our mid-teens, but I didn't really want to go to the park and sit about drinking. I knew how important the next couple of years would be.

But despite my efforts to keep up with my schoolwork, I was expected to underperform in my exams. That made me all the more determined to get ten passes, which I did. I got good enough grades that A levels were a possibility and I applied to do three. But things were changing. Now that I was sixteen, I was legally allowed to work longer days on *Emmerdale* than I had been as a child actor. As my storylines evolved, I found I was suddenly working twelve-hour days and it was less and less easy to catch up on my schoolwork afterwards. Added to that, I quickly came to understand that A levels were going to be much harder than GCSEs.

I was confused about what I should do. In the back of my mind I could still hear the voices of all the people who'd told me over the years that you 'can't *just* act'. It wasn't a viable career plan was what they told me. I had to have a back-up. If I didn't do A levels and the acting didn't work out, what would I do then?

I wrestled with that dilemma for weeks. It was hard to shake the naysaying voices, but the fact was that I had already been working as a professional actor for a decade. Yes, there was always the possibility that *Emmerdale* could come to an end at any moment – contracts only lasted a year at a time – but perhaps I was selling myself short. I'd won a Soap Award, voted for by a panel of professionals in the television industry. That had to count for something. And if all else failed, I could fall back on the skills my dad had taught me as a car mechanic. So, with my work schedule at *Emmerdale* increasing, I decided to leave school and follow my heart.

*   *   *

It was hard growing up on TV. I went through some of my biggest rites of passage as Andy on the small screen before I experienced them in real life. I think I lost my virginity on TV first! That I never went off the rails is a testament to my family, who helped me keep my feet on the ground. So much is thrown at you when you start to appear on television. They'd helped me not to believe my own hype. Now they helped me to believe in myself just enough. They helped me to see that I had to give myself the chance to follow my dream wholeheartedly.

Luckily, my decision to abandon my A levels worked out. I didn't miss school when I left for good. Not long afterwards, I also left home to share a place in Leeds with James Hooton, who played Sam Dingle. Being in Leeds made it easier to get in to work. And at seventeen, I was actually able to buy a place of my own: a two-bed mews house next to the canal.

It was a strange moment, getting the keys to my first home. I was acutely aware that my new house was worth more than Mum and Dad's and I didn't want them to think I was somehow getting above myself. I'd been raised not to be materialistic and to be comfortable with having very little. It was hard not to feel a bit guilty about doing so well financially, but Mum and Dad were nothing but pleased for me.

I went furniture shopping with Mum and she and Dad helped me to set up utility bills and things like that. I was so happy in that little house. Had I just been given a place by my parents, like some of the people I knew, I don't think I would have appreciated it anywhere near so much. But I knew how much effort had gone into paying for those four walls and that made me treat it like a palace.

It was the start of a really good period in my life. I loved having my own place and the independence it gave me. My social life was great. Because of *Emmerdale*, I got invited to all sorts of parties and openings and met some amazing people from the film and TV industries. At the same time, I had a really solid group of mates back in Royton, including my best mate, Steve Crowther, who we all just called 'Crowther'. He was always ready for a night out or a last-minute trip to the sun. There was never a dull moment. The only thing missing was someone to love …

CHAPTER EIGHT

BAD HAIR DAYS

LIZ

It was strange to be home again after a year on the road with *The Famous Five*. You would think that with such a big job my acting career had started in earnest, but the truth was I was starting from scratch. I was right back at the beginning. I didn't even have an agent. I was still getting *The Stage* and poring over the audition listings week after week; still turning up at open casting calls with hundreds of other hopefuls. I had no idea how to parlay my professional experience on *The Famous Five* into new work and a career.

Suddenly unsure whether I would be able to make a living in performing after all, I turned my attention to my schoolwork. I was studying for my GCSEs by now. Mum and Dad were keen that I had them as back-up.

I'd missed a lot of school, but I was able to get up to speed with the help of Dad. His work as a lecturer in

61

electronics meant that he was easily able to explain my maths and science homework, though it must have been frustrating for him when I complained, 'I don't need to know "why", Dad. I just need to know how!' In my head I just needed to know enough to get through my exam.

I was pleased when I got my results – a mixture of As, Bs and Cs – but they didn't help me work out what I was going to do next. I was still attending the odd audition but not getting anywhere. Michelle, meanwhile, had gone to London to try her luck as a model. I felt really left behind when she told me about her glamorous lifestyle down there. Convinced that I had to do *something* that might lead to a career of sorts, I decided to follow a crowd of my old school friends to college in Rochdale, to study hair and beauty.

I enjoyed the social aspect of college – I liked being with my friends and I loved that we got to practise hair and beauty treatments on one another. We'd have a blow-dry or free facial treatment every day. It was a really fun time. But the fact was, although I'd done some part-time work at a salon when I was much younger – I'd washed hair for £12 a day – hairdressing didn't really interest me. Nor was I very good at it. I couldn't cut it. My bobs would always turn out lopsided. I was hopeless at colouring too. I wrecked my own hair when I tried to colour it bright red, ending up with no choice but to cut it so short that everyone joked I looked like Mark Owen from Take That.

Inevitably, if a friend managed to persuade me to try out my skills on them, I would end up having to give them twenty quid so they could go and get the mess I'd made fixed by a professional!

There was lots going on in my life outside college. I was learning to drive and enjoyed going out, but in other ways I was closing down. At sixteen, I had suddenly become 'aware of myself' as teenagers do. From being the child who'd looked for any excuse to put on a show, I'd grown into a teen who just wanted to fade into the background. I no longer wanted to put myself out there. Gradually, I stopped reading *The Stage*. I stopped singing. I wouldn't even do karaoke with my mates. Though part of me was desperate to get up there and sing my heart out, I didn't want to be seen. I wanted to hide.

At the same time, I was seeing people who'd been in my acting class at the Ragged School taking their first steps into the real limelight, appearing on *Corrie* and *Emmerdale* – like Kelvin Fletcher; he had become a proper star, winning a Soap Award. I felt like I didn't know those people any more. I certainly couldn't do what they were doing. It was soul-destroying. I missed performing so much, but I just couldn't bring myself to do it. My dream had never felt so far out of reach.

I'd just turned eighteen and was coming to the end of my Rochdale college course when, on a night out in Didsbury,

I got chatting to a girl who was at university studying fashion buying. I'd never really thought about what fashion buying involved, but she made it all sound really interesting. Certainly more interesting than what I was doing.

While I had completely stopped putting myself out there as a singer or an actor, I hadn't entirely abandoned my sense of ambition. I still felt that whatever I was going to do with my life, I wanted to be really good at something. I knew 100 per cent by now that I was never going to be Hairstylist of the Year, but perhaps I could be a success in fashion buying.

Next day, I told Mum and Dad about my new friend and her degree and asked whether they thought I could go to university too. Given how important they'd considered education, they were only too pleased to hear me talk about doing more. They knew I hadn't been happy at Rochdale. Mum, who had done her law degree as a mature student, knew I wouldn't have a problem finding something suitable. With her help and Dad's, I got myself a place at Manchester University.

*   *   *

University felt very strange after school and beauty college. I was used to being told exactly what to do but at uni, from the very start, we were left to figure out more for ourselves.

I quickly realised that it was going to be hard to learn fashion marketing exclusively in a classroom. I had to be in the business to understand how it really worked. So I got myself a part-time job in the fashion department at Selfridges in the Trafford Centre, rotating between the Oasis, Kookai and Vicky Martin concessions. I told myself that, as well as helping to pay my way through my degree course, working in retail would be the perfect way to see fashion marketing in action first hand, though in reality me and my new colleagues spent a lot of time larking about and eating sweets in the store cupboards.

It was a great job. The discounts were amazing, so we all looked the business. And some of my Selfridges workmates became life-long friends, like Hayley Moynihan. I remember the first time Hayley spoke to me. I was on the Kookai concession while she was on Vicky Martin. She strode over, picked up a bag from my display and slung it over her shoulder.

'How good do I look with this bag?' she said, posing like a model.

I was struck dumb by her confidence as she held up various different Kookai outfits to herself in front of the mirror while her concession was left unattended.

Weren't we supposed to be working? I wished I had that kind of attitude. Perhaps it would rub off. Hayley was so cool. When she invited me to go out for a drink with her after we'd finished our shifts, I was thrilled.

Then there was Michelle Keegan. We clicked straight away. She was really easy to talk to and always ready for a laugh. Together with Hayley, we made a great gang. We had a brilliant time at work and out on the town afterwards and I felt my world opening up again.

I loved going out in Manchester – there was always some new bar or club to go to – but I still spent many weekends back in Royton, seeing all the friends I'd made growing up on the Thorp Farm Estate or at North Chadderton secondary school. We'd long since moved on from playing kerby. Now we all went to the same pubs and clubs. On Sunday nights, we moved around the town in a big crowd. Everyone knew the routine – first the Marsden Tavern, then Kelly's Bar (where it was a pound a pint on Sundays – though we always drank halves to look more feminine), then Revolution and Escobar, before we ended up at Liquid or Scruples where we could carry on drinking and dancing until the early hours.

The only member of our childhood gang who didn't usually pop up in Royton at some point over the weekend was Kelvin. I'd heard that he'd moved to Leeds because it was more convenient for the *Emmerdale* studios. I was in my third year at uni, working part-time in Selfridges and living in a small apartment in Royton with a friend. Kelvin had his own house, was dressing up in black tie and going to the British Soap Awards, mingling with household names. From time to time, someone saw him around and reported back that he was still the same old Kelvin, but he was living a very different life to the rest of us.

So I was surprised when I saw Kelvin in the first pub on our circuit one Sunday night. In fact, the first thought that popped into my head was, *What's he doing out? I can't believe he can get into a pub.* He still looked that young!

I was on the town with my friend Gemma, who'd just broken up with her boyfriend.

'Did you see Kelvin Fletcher?' I asked her.

'Oh yes. Fancy him, do you?' Gemma teased me.

'No way.'

The idea was ridiculous. But Kelvin was there with a bunch of lads we knew, so it was natural that we all got chatting. I remember that during the conversation Kelvin said something that made me laugh out loud. It was totally unexpected. I'd not laughed like that in a long time and I would never have imagined that Kelvin Fletcher could be the one to make it happen. But since I'd last seen Kelvin, something seemed to have changed about him. The sweet old Kelvin I knew was suddenly mysterious and cool.

Bloody hell, I thought. *Since when was Kelvin Fletcher cool?*

That night followed the routine every Sunday night in Royton followed. We went from bar to bar in the usual order before ending up at our usual club. All night, I kept on bumping into Kelvin and we'd exchange a few words until the next time. He slightly dented his newly cool image when I met him in the queue for Scruples and saw that he was wearing red-tinted sunglasses! What was he thinking? That he looked like Justin Timberlake? Even Justin couldn't really pull those sunglasses off!

Despite that fashion faux pas – and it was a big one – I found I was secretly looking out for Kelvin at each new place. But while I was looking out for him, I'd attracted someone else's attention.

At some point in the evening, a lad I didn't know had joined our crowd. He was big and muscly and really loved himself. His chat-up lines were some of the cheesiest lines I'd ever heard. I mean, I didn't even have a coat to grab … I tried to let him down gently, but he wouldn't take the hint and I couldn't seem to lose him. He stuck with our group everywhere we went. I remember grabbing Kelvin and saying, 'Kelvin, I need to escape. Can you believe he just asked me what I'd give him out of ten? I don't even know what to say to that?!'

Hard as Mr Muscle tried to win me over, I did not give in and refused to give him my phone number. However, a couple of days later, a random text popped up on my mobile from a number I didn't recognise. 'So what would you give me out of ten?' it asked.

My heart sank. I showed my friend Leanne, who was out with us that night, and she agreed it was exactly the kind of message that creepy Mr Muscle would send. I wanted to be sure, though, so she suggested that we call the mystery number from my parents' landline, having first dialled 141 so that our number wouldn't appear on the texter's phone.

We rang the mystery mobile on many occasions over the next two weeks, trying hard not to dissolve into giggles as the texter asked, 'Hello? Hello? Who is this?' while we remained completely silent. But though we called dozens of times, we were none the wiser as to who the mysterious texter might be. I didn't think it was Mr Muscle, but neither of us recognised the voice that answered our prank calls. Eventually, the game lost its appeal and we gave up on tormenting the poor bloke, without ever having worked out who he was or telling him his score out of ten!

A couple of weeks later, after a Thursday-evening shift at Selfridges, I met up with my uni friends for the student night at One Central, a Manchester club. It was a big night and we always dressed up. I was going through a phase of wearing sequinned leggings and cropped tops, with a big belt slung low around my hips. With my hair cut short, I looked every bit the fashion student.

While we were getting drinks, my friend spotted a really handsome bloke at the bar. Really handsome. And next to him? My old mate Kelvin.

Seeing Kelvin gave us the perfect excuse to go over so my friend could introduce herself to Mr Handsome. While they were chatting, Kelvin and I caught up on each other's news. It was good to see him. We talked about everything and nothing: our families, our Royton friends, his work and my studies. Then he paused, looked me deep in the eyes and said, 'Did you get my text?'

It took less than a second for the penny to drop.

'What would you give me out of ten?' I asked.

That was the one.

I laughed out loud.

'It was you!'

I didn't tell Kelvin at that point what lengths I'd gone to in my attempts to find out who sent me that message – and he didn't mention any prank calls – but I was very glad it wasn't Mr Muscle.

When we finished laughing, I realised something important. Something had clicked. Suddenly, I fancied Kelvin Fletcher. His confidence and playfulness were really attractive. While his handsome mate and my friend were busy getting to know each other better, Kelvin and I stuck together all night.

I was wearing about twenty gold chains around my neck that day and Kelvin teased me about them, calling me Mr T, in homage to the *A-Team* actor with a penchant for bling. All night long, Kelvin made me laugh and smile, and it came to me that we'd had the wrong idea about each other. We weren't the primary school kids who'd played kerby together any more (though he'd made me laugh back then too). We were meeting each other on a different level now. Why shouldn't I fancy him?

When we weren't laughing and joking, we were dancing. Gnarls Barkley's 'Crazy' was the soundtrack of the night. We had so much fun.

At closing time, as my friend and I were about to leave, Kelvin asked if he could see me again. 'Tomorrow? I'll pick you up at twelve.'

How could I refuse?

*   *   *

The following day, I was so excited. I took ages to get ready for lunch, making sure I looked just right. But at midday Kelvin was not on my doorstep as promised. Instead, he sent a text, telling me he was running twenty minutes late. I now know that Kelvin is *always* running twenty minutes late.

At twenty past midday, Kelvin arrived in a souped-up Range Rover. I had never been in such a fancy car. My own was a Ford Fiesta. I tried not to look overawed as I climbed into the passenger seat. I needn't have worried. From the minute I had my safety belt on, we were chatting. There were no awkward silences, and it wasn't just because we were childhood friends with decades of news to catch up on.

Sneaking a look at Kelvin's profile as he drove us to the restaurant, I said to myself, *This guy is the coolest.* How could I ever have thought otherwise, despite the red sunglasses? He was so polite and relaxed. It was so easy to be with him. I hoped he was feeling the same.

That day, for our first official date, Kelvin took me to a place called the Waterside in Littleborough. It was a lovely restaurant in a beautiful setting and much smarter than most of the places I usually went to. We were both still only twenty-two and the Waterside felt really grown up.

I remember that Kelvin ordered an anchovy salad. Without the anchovies. I guess that made it plain lettuce! Looking back on it now, perhaps this was the first time I thought Kelvin might be different from the other lads I knew back in Oldham who'd have pie and mash topped with four pints of lager.

We never stopped talking. It was a really nice afternoon and I would have liked to stay there until late. Unfortunately, I had to go to work. I was on a late shift at Selfridges that night. Reluctantly, we brought the date to a close and Kelvin drove me into central Manchester. When I met up with my colleagues on the shop floor, I was still buzzing from having had such a great afternoon.

It was just a couple of hours later that the phone next to the till where I was working rang. I put on my best 'Selfridges voice' to answer the call, assuming that it would be a customer. It was Kelvin. 'Do you want to meet after work?' he asked me. I didn't need to be asked twice.

We went into central Manchester for the rest of the night. We were so excited to be with one another, like a pair of teenagers. When it was time to go home, Kelvin called us a cab and made sure that it dropped me off first so he knew I'd be safe. He even walked me to my door and as we said goodnight the moment I realised I'd been hoping for came. Kelvin kissed me. It wasn't our first kiss – that had happened when we'd played spin the bottle, back when we were kids – but it was the first kiss of an entirely new era. I couldn't wait to see him again.

It was just so easy to be with Kelvin. Every spare minute we had, we started to spend together. When we weren't actually together, we were texting or talking on the phone. Our friends were really pleased that we were going out. Soon it seemed as though we'd always been together, we were that close.

We went on our first holiday together in the summer of 2006. Kelvin's good friend Ryan Thomas offered us the use of his grandma's apartment in Marbella. We spent a week there. We were really content to be just the two of us together, hanging out, chatting and joking. We'd sit on the balcony for hours, looking out at the sea, quietly comfortable in each other's presence. Sometimes we'd break the silence by saying something silly or starting a game of 'Would you rather ...?' which would result in us giggling uncontrollably until our bellies ached. We were already like two great old mates, which I suppose is what we were.

That first Christmas as a couple we went on holiday again – this time to Val Thorens in France. Kelvin pulled out all the stops to make it special. I was overwhelmed when he gave me a beautiful watch as a Christmas present. I loved it so much. I had never before been given such a gorgeous or expensive gift. It made me feel a bit embarrassed that I'd only got Kelvin some socks! And he was taking me skiing too. That was amazing. The hotel was really posh, and we were far from the usual clientele. Everyone else seemed really mature. We felt like a couple of kids in the fancy restaurant where we ordered the fondue without really knowing what it was. I don't even like cheese.

When it came to hitting the slopes, neither of us really knew what we were doing there either. I'd been skiing as a kid, but it was Kelvin's first time. We got ourselves kitted out in the ski shop, thinking we looked the business. All these years later, looking at the photos cracks me up. We don't exactly look like Olympic hopefuls. More like *Dumb and Dumber*.

But we were having the best time together. We had the same silly sense of humour. One of our favourite things to do was make prank calls. Kelvin had some brilliant ideas for those. One of the funniest was to ring up our friends, pretending we were calling from a radio station.

'It that Mr Crowther?' Kelvin would say. 'I'm ringing from Radio Manchester. You've been chosen to take part in our competition. You just have to quack a hundred times in a minute for a chance to win £250. Ready, steady, go!'

And then they'd start quacking, while Kelvin and I were practically rolling on the ground with laughter. It was amazing how many of our mates fell for it before word got out and people would say, 'It's Kelvin and Liz again, isn't it?' before they'd risk acting like ducks.

Kelvin would tease me too. When I got my first job out of university, as a buyer at an educational company that provided furniture and other equipment for schools, Kelvin told everyone who asked that I was in charge of pencils.

'She's an expert pencil buyer. Ask her anything about them,' he'd say.

You'd be amazed to find out what people want to know about pencils and pens.

*   *   *

My parents adored Kelvin. They'd loved him when he used to come and call for me after school and they loved him even more now he was my boyfriend. After we'd been together for about a year, Kelvin joined my whole family in India for my brother Daniel's wedding.

It was a big deal for Kelvin to be invited along, showing how serious our relationship was in the eyes of Mum and Dad. A couple of days into the trip, Kelvin nearly marked his card when he played another one of his pranks. One of Mum's closest friends had also joined us in India. When she went to bed earlier than the rest of the party, Kelvin called her room, pretending to be a member of the hotel staff, telling her that there was a problem with her room and she needed to come downstairs at once. Moments later, she rushed into reception in a panic still wearing her pyjamas, only to find everyone else enjoying a late-night drink. She was so embarrassed to be there in her night-clothes. Dad ticked Kelvin off, saying, 'You're just like the rest of my bloody kids.'

Luckily, what came out of the incident was that Dad really did see Kelvin like one of the family, for better or worse. And Dad sort of got his own back a few days later, when we took a daytrip to the Taj Mahal. While we were in India, Kelvin wore a lot of tight T-shirts that showed off his muscles. The local kids found this hilarious and would shout 'King Kong' at him. Anyway, back at the Taj Mahal, everyone was lining up to pose on the stone bench where that famous photograph of Princess Diana was taken back in the day. A young couple asked Kelvin for a picture. Seeing what was going on, Dad – who'd seen people ask for pictures of Kelvin thousands of times before back at home – helpfully jumped in, took the camera and arranged the young couple on the bench – with Kelvin sat in between them.

'Say cheese!' he said as the couple shared a bewildered look over Kelvin's head. They'd wanted Kelvin to *take* the picture, not be in it. They weren't *Emmerdale* fans. They had no idea who Kelvin was. I sometimes think about that couple looking back through their romantic holiday snaps, wondering who the hell the muscly guy in the sleeveless tight T-shirt was. We all still tease Dad about his mistake to this day.

Sometimes it was hard going out with a successful actor. Kelvin was earning a fortune while I was on £80 a week. When we were out, people wanted his picture and would happily shove me out of the way to get it. Drunk lads thought it was hilarious to shout, '*Coronation Street*!' when they saw him, imagining he'd be offended that they'd got the programme wrong. It could be frustrating at times, but I soon learned it was something that went with the job. Kelvin always reacted with good humour.

On a couple of occasions, Kelvin took me into work with him. On the set of *Emmerdale*, he made it all look so easy. We'd be sitting in his dressing room, messing about after raiding the vending machine seeing who could fit the most Oreos in their mouth – something silly like that – when he'd be called to the set. The moment they needed him to film a scene, Kelvin was on it. He could switch from larking about to being absolutely professional in a second. I was just so impressed.

At the same time, seeing Kelvin enjoying his work so much gave me pause for thought. He was definitely following his dream and 100 per cent loving his life. Was I?

CHAPTER NINE

OLD DREAMS AND NEW BEGINNINGS

LIZ

A couple of years after Kelvin and I started going out, I was working in fashion as planned. Unfortunately, what I'd thought would be my dream job turned out to be anything but. I was miserable. I couldn't seem to find a way to progress through the industry and I was bored. Each day felt like *Groundhog Day*. I had some amazing colleagues and of course my wardrobe was full of fantastic clothes, but I wasn't happy. Sometimes the thought that I'd be in this career for the rest of my life pushed me to tears. I remember Kelvin meeting me after work one day and finding me crying so hard that he assumed something really awful must have happened. He couldn't believe it was just my job that was making me so upset.

'I need to be doing something more creative with my time,' I suggested to Kelvin one evening. 'I think I want to get back into acting.'

But could I? Could it be that easy? Or was it just another idea that would be hard to pull off? Luckily I didn't have to spend too long questioning myself as very soon after our conversation I saw that David Johnson, our old mentor, was offering acting classes for adults on Monday evenings. Perhaps that was a good place to start.

*　　　*　　　*

I went back to the Ragged School at the age of twenty-four. More than a decade had passed since I'd last walked through those doors, and I was now an adult, but David was as intimidating as he had always been. Every two weeks, he would give us adult students a new monologue to prepare. By 'prepare' I don't just mean give it a quick read-through. You had to be really, *really* ready.

When it was your turn to give that monologue, you'd be sent into a cupboard that served as makeshift theatre wings. We're talking a proper store cupboard. You had to squeeze in alongside the cleaning equipment and the catering packs of coffee and toilet rolls. Every week I'd go into that cupboard thinking, *What on earth am I doing here? This is hopeless.*

As each student burst out of the cupboard to deliver the lines they'd learned, they had no idea how it would go, but David knew – he just knew – how well you'd prepared before you even got started. He made his judgement within a couple of seconds, barking, 'No, back in the cupboard,' more often than he let us carry on to the end of the scene.

Those classes would go on until midnight, to make sure that everyone got their time in the dark with the mop and bucket, followed by their moment in the spotlight in hope of David's praise.

It was tough. Much tougher than it had been when I was a child. But David's guidance was invaluable. All those years earlier, when Kelvin and I improvised our scene around 'intimidation', I'd had no idea what David's acting tips meant. Things like 'take it off the page' were gobbledegook back then. As an adult, I was beginning to understand. Gradually, I grew in confidence and spent more time out of the cupboard than in.

I made some good friends at David's adult classes. One of them was Charlotte Tyree. Charlotte had heard that a local amateur actors' group was putting on a production of *Rent*. She encouraged me to audition for Mimi, the female lead, an exotic dancer and drug addict. Although the show was being put on in Sheffield, which would mean a lot of travel, I was really pleased when I won the part.

The travel was normally straightforward – a ninety-minute drive – but one day, as I was on the motorway, I got caught up in the tailback from an accident. To make matters worse, my phone was on the blink, so while I could make calls, the person on the other end couldn't hear what I was saying. Kelvin, who'd heard the traffic news before I called him, was clever enough to guess what had happened.

'Go to Leeds,' he said, not knowing if I could hear him. 'Get a train from there.'

That's exactly what I did. But even the next train would leave me in danger of missing curtain-up so I had to change into my costume and do my make-up on the way. It was rush hour. As I sat at a table for four, with my big make-up bag laid out in front of me, an old couple sitting opposite watched with great interest.

'I'm in a play,' I told them as I slapped on the pan stick. '*Rent*.'

They didn't seem to know it.

'What are you?' the old man asked.

'I'm a prostitute,' I said.

I can only hope they knew I was referring to my part!

Mum and Dad came to see me every night of the run. Kelvin came whenever he could. It might have been an amateur show, but it got me back on stage and I realised that I had been kidding myself, thinking that I was ready to give up on my performing dream. I didn't want to work in fashion. The more I thought about it, the clearer it seemed. I wanted to go to drama school.

*   *   *

Facing the truth about my ambitions was a relief in some ways but a nightmare in others. Now I knew what I wanted to do, I had to ask myself how I could make it happen. I started looking at courses. Most of them were three years long and I wasn't sure I could face that. Then I found a year-long intensive course at the London School of Musical Theatre, which covered dancing, singing and acting. It was a great course – the perfect one for me – but it was so far away.

And I had to get on the course first. Telling Kelvin that I was going to London on a girls' weekend, I secretly went to an audition. I was recalled two or three more times, each time pretending to Kelvin that I was going out with friends. I didn't tell anyone – I only told my mum and dad what was really going on.

My superstitious side was what made me more secretive than normal. I think I wanted to see how far I could go before I allowed others in. Perhaps it was also because it felt like such a long shot. I already had a career that many people would die for. Was I really about to give it up and start another course? What's more, I was competing with hopeful performers straight out of school. I was twenty-eight by now. Why would the tutors choose me over them?

Shortly after the last recall, I was at work, sitting at my desk, when I had a really strong premonition that I was about to get a phone call, just as I did back when Dan

Crawford had called to tell me I was down to the last round of auditions for *The Famous Five*. I can remember it very clearly. It was July. I was writing an email to a supplier in India regarding some fabric when the phone rang. I snatched up the receiver.

'Hello?'

It was the principal of the theatre school. He was calling to offer me a place on the course starting in September. He told me he thought I would be a great asset to the school. I couldn't believe my ears.

I was over the moon, but now I had another problem. I still hadn't even told Kelvin about the auditions. We'd only just moved in together and now I was going to have to tell him that in just over a month's time I would be moving to London to go to drama school. My heart was in my mouth as I told him. What would his reaction be? Could our relationship survive?

I should have known that his reaction would be pure Kelvin. He was delighted for me.

'We'll be fine,' he said, when I told him it meant I'd have to move. 'We'll see each other whenever we can. It'll be exciting.' He finished by telling me firmly, 'Liz, you've got to do what you need to do.'

My parents were similarly pleased for me, but I could tell that the people at work thought I was crazy. They couldn't understand why I wanted to leave such a great job to go back into education. They thought I was mad to quit fashion for a career that was even more precarious, but I couldn't wait to start the next phase in my life.

*   *   *

I found a place to live in London for the duration of my studies and Kelvin helped me move down there. My first day at the college was terrifying. I was the oldest person on the course, surrounded by people an average of ten years younger. I was worried they'd think I was ancient, but my age turned out to be an advantage. Had I arrived at that college any earlier in my life, I wouldn't have been ready. Now I really knew what I wanted to do, I was committed and prepared to do whatever it took.

It was hard work from the start, but it was good to be back in the theatre world, surrounded by people keen to nurture my abilities. The great thing about the London School of Musical Theatre was that the teaching staff were not trying to turn us all into cookie-cutter versions of the big musical theatre stars. We all had a certain level of talent – that's how we'd got on the course – and we were all drilled in technique, but we were also encouraged to develop our individuality. For one, I did not have the traditional musical theatre voice. I had a husky tone, and when I sang I gave everything a distinctive pop sound. I found it difficult to hold a clean note for long enough. I thought that might be a problem, but my teachers assured me that my unique voice was a strength, not a weakness.

In many ways, that year at drama school was one of the best of my life. I realised that I was missing a part of myself when I lived in Oldham. My heart was and will always be in Oldham, but I needed a change. I'd seen Kelvin so happy in his work on *Emmerdale* and I wanted to feel the same happiness about the work that I was doing. Although sometimes it's hard to meet someone on a level when they're buzzing and achieving everything they want to and you're … well, you're just not, it's important to recognise that and to make a change, however small. For me, moving to London was the best choice I could make for myself.

Going to drama school brought me back to life. The parts of me that I'd squashed down started to blossom again and that felt good. If I'd been at all worried that going to London would be bad for my relationship with Kelvin, I soon realised the opposite was true. That Kelvin supported me every step of the way, without ever complaining that drama school was taking me away from him, was proof of how strong we were as a couple. What's more, as I grew in confidence, I was beginning to feel more like the girl I was supposed to be. I found my smile again.

CHAPTER TEN

THE NEED FOR SPEED

KELVIN

Liz wasn't the only one exploring new dreams – or rather old dreams – at around that time.

Ever since I was a little kid, I've been fascinated by anything and everything to do with cars and motorbikes. My parents even claim that while most small kids go to bed with a teddy bear, I preferred to go to bed with a spanner tucked under my pillow. From the moment I got on the PW50 miniature motorbike Mum and Dad bought me for Christmas when I was four, I've had a need for speed.

When we were living in Derker, Dad finished his day job at around seven each day, but as soon as he got home he started again – fixing people's cars on the street outside our house. I loved watching him getting those broken-down cars going again and wanted to be a mechanic just like him. When I was old enough, Dad put me to work as his assistant, getting me to pass him his tools while he worked. I itched to be allowed to fix those cars myself, but, quite rightly, he wouldn't let me do anything serious until much, much later. Dad was a perfectionist. He still is.

He was also a great driver. Sometimes, we would take his car somewhere quiet – like an empty supermarket car park – and he'd let me sit on his knees and turn the steering wheel while he did the pedals and gears. Later, when I was too big to sit in his lap, he'd let me change the gears while he was in the driving seat. I couldn't wait to do it all myself. The day when I turned seventeen and would be able to get my driving licence couldn't come quickly enough.

I didn't have to wait that long to get behind the wheel, though. One day, when I was about fourteen, Mum and I had a row, and in a fit of anger I picked up her car keys and stormed out of the house. Mum's red Vauxhall Corsa was on the drive. I got into it and started the engine. I revved it loudly, just to annoy her, but she didn't come out of the house. I revved it again. Still nothing. She stayed inside. She must have decided she wasn't going to rise to the provocation.

Right, I thought. *I'm going to drive this thing.*

How hard could it be? For years I'd watched Mum and Dad drive. I knew the theory. All I had to do was put the clutch in, put the car into gear and let the clutch out slowly while pressing the accelerator down. It had always been a massive ambition of mine to drive on my own. Why not start now?

I put the car into reverse gear and drove it backwards off the drive. It was a diesel car but somehow I managed to get it going without stalling. Once on the road, I put it into first and edged forward. A little faster and I was in second. Third gear. Fourth. I could do it! I was driving!

I drove all around the estate, going past our house a number of times. Eventually, after twenty minutes without hearing me revving the engine, Mum came out of the house to find that her car had disappeared. When I saw her standing at the top of the drive as I drove back up the street I expected her to be furious. I was sure I was due a serious telling-off. Instead, Mum had a look of confusion on her face, which quickly turned to a beam of pride when she realised it was me behind the wheel.

'You're driving!' she said. 'Kelvin! You're driving!'

When Dad came back from work she made me show him what I could do, and he seemed as impressed as Mum was. Whatever it was we'd argued about earlier that day was completely forgotten, though it would be a number of years before I was allowed to get hold of Mum's car keys again.

As soon as I was old enough to get a moped, I bought a Vespa with my *Emmerdale* wages. I loved *Quadrophenia*, the film about Sixties mods and rockers. Since I'd first seen it, I'd been an aspiring mod.

Around the same time, we moved to a bigger house with a bigger drive. Now that he had the space, Dad bought a clapped-out Mini and every night when he came home from work (he had an office job in transport in those days) he'd change out of his suit and I would help him fix the car.

Fixing up that car together was a real father–son bonding experience. Dad has always been a man of few words, but those moments when we worked on the Mini were precious. Though I knew he was proud of the way I was getting on as an actor, there was still a part of me that wanted to impress him by mastering the mechanical skills he showed me. The first time I fitted a carburettor and brake pads on my own, I felt as proud as I did when I first started getting acting jobs.

Dad and I also bonded by watching motorsports together. Not so much Formula One – though of course we were glued to the British Grand Prix and I was a big fan of Nigel Mansell – but the British Touring Car Championship, motorbikes and truck racing. One of the perks of being on *Emmerdale* was that I got to meet celebrities from other walks of life at parties and charity events.

I was lucky enough to meet several of my motorsports heroes, including Johnny Ray, Leon Haslam, John McGuinness and Marco Melandri.

The minute I turned seventeen, I was ready to take my driving test and graduate from two wheels to four. I booked the earliest possible test slot and bought myself a VW Polo GTI in readiness.

I was a confident driver and thought that passing my test would be a formality, so you can imagine how disappointed I was to learn that I hadn't passed. I couldn't believe it. I was heartbroken. The worst of it was, I felt I'd let my dad down. It was no consolation to me to hear people – including Dad himself, despite the fact he passed his first test – say, 'All the best drivers fail first time.' I knew that even Nigel Mansell had failed his first test, but I didn't care. The thought of passing mine had meant so much to me.

I passed second time, thank goodness, and from then on I was unstoppable.

* * *

Around the same time as Liz decided she wanted to go to drama school, Dad and I went to a big motorsports exhibition. While we were there, we got chatting to a guy who raced Minis. He took us through all the steps involved in getting a racing licence – getting a medical, things like that. It sounded much less difficult than we'd imagined. Both Dad and I had always wanted to try racing ourselves. Perhaps we should give it a go …

A couple of weeks later, I had my racing licence and Dad came with me to look for a car. A Mini was a good car to start with. Thanks to Dad's knowledge and the skills he'd shared with me, we could buy something that other people might have thought needed too much work to get it race-ready. We signed up to a race at Oulton Park to give ourselves a deadline.

Motor racing isn't exactly a cheap hobby, but there are ways to do it on a budget. Dad and I split the costs, and to save on hotels we bought a two-man tent to use on race weekends. We didn't play golf. We didn't go on lads' holidays. Racing that Mini was our one indulgence. It was cheaper than the average lads' weekend.

One of the things Dad and I enjoyed most about our first steps in the sport was the sense of community to be found at the racetrack. The drivers came from all walks of life. They were men and women, young and old. Though we'd be competing against each other on the track, off track everyone was really friendly.

In my very first race, I started on pole and came second out of thirty-odd drivers. It was a great result and really gave me the bug. The only problem was, strictly speaking, I wasn't supposed to be racing a car at all.

My *Emmerdale* contract specified that there were certain things I was not allowed to do – like skiing, for example, in case I broke my leg or something like that and wasn't able to shoot my scenes. It was understandable and I didn't mind too much. I'd realised after that trip to Val Thorens with Liz that I didn't really want to ski anyway. For the most part, I was happy to keep my sporting exploits to a bit of charity football. But motor racing was another one of those things that would probably have been considered too dangerous, and that I did want to do.

To begin with, I kept my racing quiet. But it wasn't long before I got a place in a championship that was going to be broadcast on ITV. There was no way I would be able to keep that under the radar. I had to tell the producers. I argued for a special dispensation. They still didn't want me to do it, so I suggested that if I wasn't able to race, I'd have to think about leaving the show. I was allowed to race, so long as I took out special insurance.

Emmerdale took up 100 per cent of my working life, but the pattern of recording allowed me to have other interests. It also gave me the time to prepare for what I'd do if *Emmerdale* was no longer there. The job felt secure but that was an illusion. Your character could be written out at any moment. If you were killed off then that was definitely that. By the time I was in my late teens, I'd seen many actors come and go, so I'd decided I never wanted to be in a position where being suddenly axed from the show was going to be a big problem for me. I anticipated and expected that one day it would happen. Investigating other options kept me sane. Motor sports was one of those options.

Since those early days in the Mini, I've worked my way up through the classes and now even have my own team – Paddock Motorsport – which I started with my friend Martin Plowman, winner of the 24 Hours of Le Mans. We've had some big successes and have been able to give several younger drivers their first break when they were starting out.

After acting, my big passion was always cars and I feel lucky to have been able to fulfil my driving dreams – though these days you're just as likely to find me behind the wheel of a tractor. I must be the only farmer in the Peak District who takes his tractor through the car wash. Still, I think Liz appreciated the effort when I gave her a lift to Alderley Edge to get her nails done in the cabin of my new Valtra.

CHAPTER ELEVEN

LOSING MY VOICE, FINDING MY VOCATION

LIZ

Going to drama school gave me a new sense of purpose, but it wasn't all smooth sailing. As I learned more about singing technique, I started to worry about my voice. I became obsessed with it, with improving and protecting it. Great singers make it look easy to get up there and sing so that a whole theatre can hear, but it really isn't easy at all. While my younger classmates were making the most of being students in London, I didn't dare go out in the evenings, in case I lost my voice the following day. I hadn't come all this way to throw away my chances on a big night out.

My worries weren't unfounded. Halfway through the course, the school's principal sent me to a doctor who specialised in the vocal cords. The doctor told me that my voice was being affected by muscle tension in my neck due to my straining to hit certain notes.

After this diagnosis, I became incredibly stressed about saving my vocal cords from further harm. I started speaking very quietly. It was a nightmare to be out with friends in loud restaurants. I just wanted to stay at home in the quiet, waiting for the moment when I would really need to use my voice, such as during my final term, when I would be performing for agents in the hope that one of them would take me on and help me build my post-drama-school career.

Each of us students rehearsed a number of songs for the agent auditions. Because I didn't have the traditional soprano voice that many of the girls on the course had, I'd chosen to rehearse a number of belters, including Aerosmith's 'I Don't Want to Miss a Thing'. It just so happened that after long days spent listening to the usual musical theatre showstoppers, the belters were the songs the agents really wanted to hear.

My choice of music worked. I received a number of offers of representation. I signed with my preferred agent at a great agency and allowed myself to relax for a moment and think, *This is it. I'm on my way.*

But fate, as always, had other ideas.

*   *   *

During the process of auditioning, singing those belters every day, I'd noticed that my voice seemed to be getting deeper and deeper, huskier than ever. Every day I downed as much water as I could, thinking that would help. It didn't make much difference, but I couldn't think what else to do. I was massively relieved when the auditions were over. Now I could give my voice a rest. Except I didn't.

The college was putting on an end-of-year show on a barge on the River Thames. It was going to be a real celebration and of course I wanted to take part, singing one of my favourite songs – another belter.

It was a great night and I thought I'd got away with it. But right after the show ended, something weird happened. I felt an odd popping sensation in my throat and knew instinctively that something had gone very badly wrong.

As soon as I was able, I went back to the throat doctor. He examined me and delivered the bad news. This time I had suffered an actual haemorrhage of the vocal cords – a common problem for singers.

'What can I do?' I asked him.

'You'll have to have an operation,' he said.

As I left his surgery, I felt like such a failure. I'd finally found the balls to go back to drama school. I'd given it my all. I'd signed with a great agent and now this? After taking so much care, I'd still managed to ruin my voice. I was devastated.

At first I couldn't bring myself to tell anyone except my parents and Kelvin. Fortunately, when I did get round to telling him, my new agent was understanding. Mum came with me to hospital for the operation and it went well. Much more difficult was the fact that, to recover, I was going to need a period of six weeks' strict vocal rest. This applied not only to singing but to speaking too. I had to remain completely silent 24/7. After that came a period when I would have to rebuild my voice by talking for just ten minutes a day. It felt like the end of the world.

I moved back home with Kelvin. He understood what I had to do to save my voice, but it wasn't easy. Kelvin and I could sit quietly together but there were other people popping round all the time. We still had so many friends in Royton and it was agony not being able to chat to them when they came over. I had to hide myself away in the bedroom to resist the urge to join in.

It was a really weird time. Our house had been my sanctuary but while I was recovering from my op and thinking about the next step – if there was a next step – it felt like my prison. Without my voice, who was I? When Kelvin was at work, I would sit in silence, just waiting for him to come home so I could spend my ten-minutes' speaking allowance on him. But no matter how much love and support Kelvin gave me, ultimately, when it came to getting my voice back, I was the only one who could do it.

I had to follow the doctor's orders and that meant no or only limited talking. For someone who always has something to say, I found this the hardest. I felt a growing distance between me and the outside world. I was beginning to turn in on myself, living inside my own head, unable to speak and share my thoughts. Everything was so quiet, but my mind was anything but. Would I find my voice again? Would I ever be able to sing again? Was this the end of my dreams? So many questions!

The entire process of healing from the operation took six months but it had a permanent impact. The experience had left me traumatised. I was terrified of using my voice on stage again. How would I ever find the confidence to hit the big notes, knowing that I might have another haemorrhage at any time? I just couldn't face the idea of losing my voice again and spending another half-year in virtual silence.

I began to realise that I couldn't spend my days worrying about whether the next song might be the one that broke my voice. Maintaining a great singing voice required too much commitment at the expense of everything else. It was no way to live. Reluctantly, sadly, I realised that at least in the short term I had to let my dream of being a professional musical theatre performer go.

It was hard, but I could still act and dance. I taught dance classes for primary schools to make some money while I waited for auditions. To begin with, I didn't tell Kelvin when those auditions came in. I was as secretive as I had been when I first applied to drama school, because I still felt it would make things easier if things didn't work out as I hoped. If no one knew I'd gone for a job, then no one would know that I hadn't got it. I now know that many actors approach auditions in the same way, knowing that nine times out of ten they won't get the part. You don't want to have to be continually telling people about the times you didn't get chosen.

Not long after I started looking for work again, I was invited to audition for *Coronation Street*. It was a massive deal. When I got three recalls, I started to think this might be it. If I got a regular part on *Corrie*, it would change everything and make all the pain of losing my voice worthwhile. Finally, I was invited for a screen test with a veteran star of the show. I still hadn't told Kelvin. Whenever he walked in on me at home while I was looking at the audition script, I'd quickly hide it away. I was incredibly nervous, not least because this particular star's reputation preceded them.

Indeed, the veteran star put me through my paces, and though I did my best and the casting director told me I'd done really well, I didn't get the part. In retrospect, that was a good thing. And not long afterwards, I did land my first TV job: a part in Kay Mellor's new drama, *In the Club*.

By now not telling Kelvin I had an audition was part of my superstition, but this time I took a big risk and told him, because the offices where the auditions were being held were right next to the *Emmerdale* set!

I'd always been a big fan of Kay Mellor and wanted to work with her so much. She really was the queen of northern drama. I loved *Band of Gold*, *Playing the Field* and *Fat Friends*. Kay was an actor and director as well as a writer, so she knew how to get the best out of her team.

Sometimes, you can tell the minute you walk into an audition whether or not you've got a chance of getting the job. After my audition for *In the Club*, the casting director seemed pleased and Kay Mellor told me, 'I can't wait to watch that film back.' It was a positive sign that did me the world of good after the disappointment of my *Corrie* screen test. That audition convinced me I could still do it. And I did get the part.

In The Club followed six couples who meet at a parent craft class during pregnancy. The first read-through was attended by the whole cast. There were loads of faces I recognised, and it was hard not to be star-struck.

My scenes were with Hermione Norris, who was already a big star thanks to *Cold Feet*. Despite her fame, she was really easy to work with, treating everyone on a level. Still, kind as Hermione was, I was having to act in two dimensions. I was acting the part Kay Mellor had written, but I was also having to act the part of a seasoned performer. I was literally learning on the job but couldn't let anyone see that. All around me, the crew were saying things that sounded like a different language, yet I couldn't ask for explanations. That would give me away as the newbie I was.

I didn't even ask Kelvin for his take on things when I got back home at night, which I now know he found really hard to accept. He would have loved to be able to help me. But every acting job is different, and it wasn't necessarily the case that Kelvin's experiences on *Emmerdale* would be relevant to *In the Club*. I wanted to find my own way.

Though it was hard work, *In the Club* was a great way to start my television career and definitely helped me get over the disappointment of realising that I wouldn't be starring in a West End musical. These days, my voice is completely fine, but I believe that things happen for a reason and that I was meant to go through that tough time.

Through David Shaw, the casting director who together with Kay Mellor gave me my break, I'd come to learn about the Jungian archetypes, and the related theory that if you're on the wrong path in life, you'll be shaken off it by chaos so that you can find the right path in the aftermath. This process happens every seven years or so. I believe that's what happened to me when I turned twenty-eight.

Because I'd been so scared of using my singing voice again, I put more of my focus into acting and into voice-overs. I realised that my speaking voice – Oldham accent and all – is a real asset. I've been the voice of the Co-op and Google. Being a voiceover artist is the perfect career for me. I've also realised how hard it would have been to combine working in live musical theatre with mother-hood.

That said, never say never. Of course I could never really give up singing. With my friend Michelle, I've sung in clubs and at weddings. And I sing all the time with the children at home. They love it when we sing together.

Looking back, I can see that the highs and lows we experience in life always bring us to the right place in the end. Hopefully, there's still a moment in my future when I take to a West End stage.

CHAPTER TWELVE

IF YOU HAVEN''T PROPOSED,
I'M OFF!

KELVIN

I knew how hard it was for Liz to give up on her musical theatre dream, so I was really happy and proud when she started getting good acting jobs. Now that she was back in Oldham, our relationship was more solid than ever and we supported each other through everything.

By 2014, we'd been together for the best part of eight years. We had a house together. Our friends and families were intertwined. There was no doubt that we were serious about each other – we had been from the very beginning. Lately we'd started talking about the possibility of having a family. I'd always known that Liz was driven to be successful and was not going to be happy as a housewife, staying at home and making the tea, but we thought that if we had children we would be able to make it work so that both of us could still have our careers.

Liz was adamant about one thing, however. She was not going to have a baby unless we were married. No way. If we had children, it would be after we got wed. She wanted them to have that security. So at the beginning of the year in which she was going to turn thirty, Liz gave me an ultimatum.

'If you've not proposed by the end of the year, I'm off.'

She delivered it in a light-hearted sort of way, but I sensed that she really wasn't joking. What was I going to do?

*   *   *

The soap life is a great life. I'd have to work really hard in intensive bursts when my character had an important storyline, but then I might get a whole week off. When that happened, Liz and I made the most of it, often going away for mini-breaks. In November 2014, I had a spare week so we decided to go to Anglesey, taking up a holiday that we'd won in a charity auction earlier in the year.

November in Wales probably wasn't Liz's idea of a glamorous holiday, but we were going to be staying in a really lovely cottage on the beach. When we arrived it was cold, so we laid a fire in the fireplace and soon had it lit. But just as the house was starting to warm up, I realised I'd messed up my timing.

'We've got to go out,' I said. 'We've got to see the sunset, have a walk.'

'What? A walk?' Liz said. 'But we've only just got the fire going.'

'We can leave it.'

'We can't leave it. What if the house catches fire while we're out?'

Liz had a point.

'We'll go for a walk tomorrow,' she said.

She didn't know why that wouldn't work for me.

'Come on,' I said, pulling her up from the sofa. 'How often do you get to see the sunset over Anglesey?'

Liz narrowed her eyes at me, as though she knew something was up.

'You're crazy,' she said. 'But if you really want to go, we've got to put the fire out first.'

Ridiculously, we picked up some of the bigger logs using oven gloves and carried them, smouldering, to the water's edge. Once that was done, we set off on our stroll.

LIZ

As soon as we got to Anglesey Kelvin didn't quite seem like himself and I wondered what was wrong. Why was he insisting on going for a walk when it was so cold and we'd just got the cottage warmed up? He was unusually quiet as we headed towards the beach and the weak winter sun began to set. I knew something was up, but I didn't have a clue what it was.

Two years before, Kelvin had taken me to Paris for my birthday. It was an amazing weekend. Really romantic. So when Kelvin insisted on that trip that we go for a walk to see the Eiffel Tower at midnight, I dared to think that he might be about to get down on one knee. It didn't happen – we just looked at the famous tower, then went back to our hotel – and the following day, as we were walking along the River Seine, Kelvin actually said to me, 'By the way, I'm not going to propose.' Though of course I laughed it off and tried to act as if I'd never dreamed of such a thing, I was seriously annoyed. I didn't say a word to him for at least half an hour. I remember standing in front of Notre Dame, inwardly fuming. I'm not sure if Kelvin even noticed that I wasn't speaking.

When we got back to the UK, I told my Mum and Dad what had happened. 'He's bloody freewheeling!' Dad said of Kelvin. My parents loved Kelvin like one of their own, but Dad was clear that he thought it was high time my boyfriend put a ring on my finger. That must have been in my mind when I gave Kelvin that ultimatum at the beginning of the year, though by now we were in November and there was still no sign that Kelvin felt moved to act on it.

I'm sure I was already planning to quietly drop my dead-line and give him another twelve months!

A proposal definitely wasn't on the cards for this week-end. Anglesey wasn't Paris, New York or Rome. Not one of the obvious romantic destinations. There was no point getting my hopes up.

So we went for our walk on the beach, with me thinking only of getting back in the warm. Then suddenly I realised that Kelvin had stopped walking, leaving me striding on ahead chatting to thin air.

'Kelvin …'

I turned around to find that he had gone down on one knee in the sand.

I put my hands to my mouth in disbelief as Kelvin began a speech that he must have been practising for weeks. He spoke so beautifully about what I meant to him, by the time he pulled a ring out of his pocket and asked the question I had been waiting for, I was ready to burst with delight.

'Yes!' I shrieked. 'Yes! Yes! Yes!'

This was why he'd been so determined that we get out of the house. This was what he had been planning. Kelvin presented me with the perfect diamond. We kissed and hugged and danced on the beach. It wasn't Paris but I was glad about that. I could not have dreamed of a better proposal.

Once we'd finished jumping up and down with joy, we took pictures on the beach to send to our loved ones. I quickly found out that I was the only one who'd been surprised by Kelvin's proposal that day. He had been planning that evening for ten months – from pretty much the day I gave him my ultimatum. He'd had the ring made specially, sourcing the diamond himself. He'd travelled to Anglesey in secret to recce the exact spot where he would pop the question and researched what time the sun would set. He'd written to my parents to ask their permission to marry me, which of course they had happily given. They knew that Kelvin and I shared a great mutual respect and love and we all joked that he was already their 'golden child'. They were just relieved I'd said 'yes'!

Kelvin and I wandered back to the cottage hand in hand, talking excitedly about the wonderful years that lay ahead. But as we turned a corner, we saw that the sky was a really strange shade of orange. Not a romantic sunset orange. The kind of orange you only get from flames …

'Kelvin! The cottage is on fire! I knew it! I knew this would happen!'

We ran the rest of the way, convinced that we'd not been as thorough at putting out the logs as we thought. We were going to be in so much trouble. It was only as we got to the gate, our sides aching from the run, that we realised the flames were not coming from our cottage at all, but from the house behind it. The owner was having a late-night bonfire in his garden.

Reassured that our cottage was safe, we got ready to go out to dinner at a nearby restaurant. Kelvin was fizzing with excitement and told anyone who made eye contact that he and I were getting married. He wanted to share our joy with everyone. I was more than happy with that.

Getting ready for bed that night, I looked at my brand-new fiancé, my old friend Kelvin, and I felt my heart fill with happiness. I knew I'd made exactly the right choice.

KELVIN

Though I'd been pretty sure that Liz would say yes, especially since she'd given me that ultimatum, it was a massive relief to have that ring on her finger. We got engaged on 28 November 2014. By the time we left Anglesey for home, we'd set our wedding date for exactly a year later.

When we started talking about it, Liz initially wanted a low-key wedding, but I had other ideas. I wanted our wedding to be a proper party. I remembered my dad telling me that there are two times in your life when you can get everyone you love together in the same room. 'One of those times is your wedding,' he said. 'The second time, you won't be there to see it …'

With Dad's words in my mind, I persuaded Liz that we should pull out all the stops. We had to make our wedding day one to remember.

*   *   *

Over the next couple of weeks, we visited hundreds of wedding fayres in our search for the perfect venue. We looked at all the country-house hotels around the North. They were all great but none of them seemed quite right for us. We wanted something a bit different, something that everyone would remember – not just another cookie-cutter do. All the same, I was surprised when Liz suggested that we look at a venue in London. We were northerners. Our family and friends were in the North. What did we want to get married in London for?

Liz persisted. She'd heard about a place called One Mayfair, a deconsecrated nineteenth-century church right in the centre of the city. She persuaded me to go and have a look. There was no doubt it was amazing. The nave of the church had been converted into one big event space that had hosted not only weddings but film premieres.

As we were shown around, we realised that One Mayfair had everything we were looking for. It had drama and flair. It had atmosphere and class. We could imagine ourselves putting on a really spectacular winter wedding there. It was also a venue where we could imagine our family and friends having a really good time. This opportunity to show how much we appreciated everyone in our lives was not to be missed.

LIZ

With the venue, the entertainment and the catering coming together nicely, I went shopping for my dress with Mum and Grandma. Grandma, who had recently turned ninety, wanted to buy my dress as her wedding gift to me. It meant a great deal to me to have Grandma with me and Mum on such an important day.

We went to a shop in Uppermill, not far from where Kelvin and I had had our first date. It was an emotional moment, finding the dress of my dreams and seeing Mum's and Grandma's smiles as I stepped out of the changing room. I chose a strapless lace fishtail gown embellished with Swarovski crystals. It was traditional but ornate. I matched it with Charlotte Olympia shoes, but it was the veil that really made the outfit. I couldn't wait to walk down the aisle and see Kelvin's face. But before that big moment arrived, Kelvin and I had our hen and stag parties to get through.

As soon as I knew I was getting married, I knew exactly what I would do for my hen party. In July, I gathered thirty girlfriends and took them to Marbella.

It was full-on 'Marbs' from the moment we touched down at the airport. I would say I'm not really a party animal, but there was no way I was going to spend my 'last night of freedom' sat in a Wetherspoon's. We needed sun and sangria. We wanted to dress to the max. We were going to be out all night, every night.

All my favourite women were there with me. Michelle Keegan was a last-minute addition, having thought she wouldn't be able to make it due to work commitments. It was too late to get her a room of her own, but she said she would be happy sleeping on the floor if she had to. As it was, we were in an apartment with a kitchenette and Michelle curled up on the bench under the breakfast bar each night. She assured us that it was actually really, really comfy.

We were determined to go all out for the whole week-end, so to pay for our drinks, we each put £200 into a kitty. It seemed like that should be more than enough for a few big nights, but we soon learned that the Marbella life can be *really* expensive. There were clubs where, to get a table, you had to be spending thousands from the minute you sat down.

At one club, we were invited a join a table of high-rollers – one of my hens knew one of the guys from back home. They'd just ordered a bottle of vodka that cost the best part of five grand. For that, the vodka was delivered to the table by a parade of fabulous-looking girls in bikinis accompanied by music and fireworks. The whole club stopped to watch.

We were really excited to be a part of it. We were dancing and cheering and loving the party atmosphere. The guys who bought the vodka were being really generous. My friend Leanne McCall was the first to be offered a drink from the five-grand bottle. She picked it up, poured herself a shot and … the bottle slipped straight through her hands and smashed on the floor. We danced off pretty quickly after that and as we stumbled out of the club into the daylight, we realised it was eight in the morning.

My bridesmaids were brilliant. As we sat by the pool, they handed out 'Liz masks' for a brilliant group photo. They had also organised a surprise lesson at a dance school, where we spent a couple of hours learning a routine to Beyoncé's 'Single Ladies (Put a Ring on It)'. It was brilliant fun, even if the Beyoncé leotard they'd bought for me specially was so small it was like wearing a cheese wire. Even my best friend Michelle, who had just found out that she was pregnant, got up and danced. It was one of the best weekends ever.

KELVIN

I had three stag dos. I mean, you only have a stag do once, right? Or three times … The first was in Tenerife.

There were fifty of us. I know it's traditional for the best man to organise the stag do but I wanted it to be perfect, so I took control – much to my mates' amusement. I arranged the flights and the hotel and told my stags that the last one to pay his share for the weekend would get a special surprise. Actually, all of them got a surprise. I'd decided I was going to wear my *Emmerdale* overalls for the flight out to the island, while my mates would be dressed up as cows. I was Andy Sugden, taking my cows on holiday. And a pig. The last man to pay up had to wear a pig costume. That man was James Hooton. He took it really well.

Beneath the matching cow costumes, my stag party was an eclectic mix. Crowther had to be there. No party was complete without him. I took my brothers, of course. I also took Dad and Grandad. There were friends from my childhood and friends from my *Emmerdale* days. There were friends from motor racing. These were people whose paths wouldn't normally have crossed, but as I herded them through the airport I knew we were going to have a great weekend and we did, making memories that I definitely can't share here.

My second stag do was closer to home. Once again, I organised everything. I didn't just want a boozy send-off into married life; I wanted to show my friends how much they meant to me and give them an experience they'd never forget.

This time I rented a field in Cumbria for three days and filled it with tents. There were seventy of us – once again, from all walks of life. Some were rich, some weren't. Some were old, some were young. Camping was a leveller. We created a safe space full of love around the campfire. By the end of the weekend, we'd made such tight bonds, even the friends I'd expected to be more reserved were ready to look into the flames and share hopes and fears that they'd never shared before. I was so happy to see all these men I loved making connections that I hoped would last them a life-time.

Stag do number three was a traditional night out in Manchester. It was low-key but just as special as the other two big events. I was humbled and grateful that so many people made such an effort to celebrate with me. And by the end of that night, I thought I was just about ready to get married.

CHAPTER THIRTEEN

WE'RE GETTING MARRIED IN THE MORNING

KELVIN

The night before the wedding, I travelled down to London to the Grosvenor House hotel, where my groomsmen and I were going to be staying. Liz was going to be in the same hotel, though we planned to make sure we didn't see each other before she arrived at the ceremony the following day. That didn't stop me from getting the hotel to let me into the room she would be staying in to leave her a wedding-eve gift.

I'd been planning something special for a long time, though perhaps not consciously to begin with. At the end of our lunch at the Waterside at Littleborough, back in 2006, I'd picked up the restaurant's business card and tucked it into my wallet. It had stayed there ever since. Occasionally, I took it out and looked at it, and remembered that long-ago afternoon by the river and how wonderful it had been. Nearly ten years on, Liz and I were about to get married and I still had that card. Though now it was battered and worn at the edges, it was really precious to me.

Everyone kept telling me that I needed to buy Liz a present to open the night before the wedding. My guess is that they thought I'd go for something flash, something encrusted with diamonds or a pair of expensive designer shoes. I was in a taxi, driving across London when my gran called to remind me about the importance of the pre-wedding gift.

'Don't forget, Kelvin!'

I got the taxi driver to pull over outside Hotel Chocolat and rushed in to buy a box of posh chocs. Luckily it was right next to a flower stand, so I grabbed a beautiful bouquet too. But that wasn't the gift. The important thing was the business card from the Waterside. I wanted to show Liz that some part of me had always known that she and I would end up together. On the back, I wrote a special secret message.

When I let them in on the plan, the staff at the Grosvenor let me into Liz's room so I could set up my surprise. I stood at the door and imagined Liz arriving. What would she see first? Where was the best place to put the flowers, the chocolates and the all-important card so she would clock them as soon as she walked in? I set the scene as carefully as any theatre director, then slipped out and closed the door behind me, imagining Liz's face as she found that old business card – I wished that I could be there to see it – then I went down to the bar to meet my groomsmen.

LIZ

I arrived at the Grosvenor a couple of hours after Kelvin. I'd driven down from Oldham with Mum in a car stuffed to the roof with everything we needed for the wedding. My dress seemed to take up half the space. So, when I walked into the room that Kelvin had booked for me, I was laden with bags and boxes and I was a bundle of stress from the drive and the thought of everything that would be happening the next day.

Planning the wedding had been like planning a show in which Kelvin and I were not only lead actors but also the producers and directors, the set designers and the catering team … So many things had to come together at once and in a perfect way. Foremost in my mind was that I needed to find somewhere to hang up my dress and all of the brides-maids' dresses, which might have got creased on the drive.

I was so focused on getting all my luggage into the room that I didn't have a chance to take in my surroundings, how lovely the room Kelvin had chosen for me was or anything else about it. I was too busy unzipping garment bags and shaking out layers of tulle and sequins. I was looking for somewhere to put my wedding heels. I was trying to find a space for my cases ...

I must have been in the room for at least half an hour before I finally sat down on the bed and kicked off my shoes. That's when I noticed the flowers and the chocolates. It was another few seconds before I registered the card.

To anyone else, it was just a tattered old business card, but to me it meant the world. When I held it in my hands, it took me right back to that afternoon at the Waterside. I could almost feel the same butterflies in my stomach that I'd felt that day, as I tried to act cool even though I already knew I was falling in love. It meant so much that Kelvin had kept that card, which showed that he'd felt the same way too.

Don't get me wrong, I wouldn't have said no to a diamond necklace, but I still wanted to throw my arms around Kelvin and tell him how much that simple gesture meant to me. However, we had promised that we wouldn't call each other that day. Not being able to speak to him made it all the more romantic. Having to stay apart made me even more excited about getting married to him in front of all the special people in our lives.

*   *   *

Next morning, the bridesmaids gathered together in my room to get ready. I had four: my best friend, Michelle, who had been there at the very beginning when Kelvin walked into Thorp Primary; Hayley Moynihan, who I'd met when we worked at Selfridges; my Ragged School friend Charlotte Tyree, and Leanne Morgan, another friend from school. Hayley had flown in from Oman where she was working as an English teacher.

It's a tricky business, choosing bridesmaids' dresses when you've got four very different women to suit – especially when one of them is heavily pregnant, as Michelle was by November – but I think I did a good job. I avoided any flouncy horrors and instead dressed my bridesmaids in gold sequinned gowns that made them look as glamorous as the Supremes. We were all going to carry circular bouquets of white roses and hydrangeas.

I'd decided against hiring a professional make-up artist. In the run-up to the wedding I'd done loads of hair and make-up trials with various different people and never liked the outcome. They always left me looking not quite myself, with more contouring than an Ordnance Survey map of the Peak District, and on this day of all days I wanted to look like me. I was really lucky that Leanne is a brilliant hairdresser and she offered to do the bridesmaids' hair, while my friend Claire, who also used to be a hair-dresser, burst into my room in the morning, curling tongues pointing at me, saying firmly, 'There's absolutely no way you're doing your own hair on your wedding day, Liz.' It was hard to argue with that, or the curlers, so she plugged them in and started working her magic.

Meanwhile my phone was buzzing non-stop with messages from friends we'd be seeing later. Michelle, who'd been married to her own sweetheart, Mark Wright, for just six months, sent me a few pieces of advice, gleaned from her own big day. She told me to make sure someone got a photo of Dad seeing me in my dress for the first time, to get his reaction. She also reminded me to take in every second and make the most of it, because no matter how long the run-up seemed, the day itself would go by in a flash.

Mum and Dad arrived a little later, filling the room with jokes and laughter as my bridesmaids and I put the last-minute touches to our make-up and hair. I stepped into my Charlotte Olympia wedding shoes and we all posed for a few group photographs. Then, finally, we were ready.

Mum and my bridesmaids went on ahead to the venue, leaving Dad and I to share a special moment as we waited for the bridal car. It was hard to keep from crying when Dad told me how proud he felt.

As the moment to leave for One Mayfair grew closer, I got a bit quiet. It struck me just how big a deal this really was. I was getting ready for my wedding. Kelvin and I were making a commitment that would last for the rest of our lives. I'd known for a long time that there was no one else I'd rather be with, but this was a different level. The magnitude and gravity of the moment sank in. We were actually doing it. Me and Kelvin were getting married.

As I walked into One Mayfair with Dad, a violin and a piano were playing Des'ree's 'I'm Kissing You'. Dad held my arm tightly and whispered in my ear, 'Walk slow, Elizabeth. You'll never have this moment again.' So I slowed my pace to match Dad's, enjoying every second, gazing around the room and seeing in every chair someone who was close to my heart. Lit by candles on crystal stands and chandeliers, the whole church glittered and shimmered like something out of a fairy tale. And there, at the top of the aisle, was the love of my life …

CHAPTER FOURTEEN

WE DO

KELVIN

That moment, standing at the front of the church, waiting for the bride to arrive, is nerve-wracking for any groom, but on the day that Liz and I were to get married, I'd never have believed I could feel so nervous. Not least because Liz didn't appear on time.

I'd heard it was traditional for the bride to be a little late, but after three-quarters of an hour, I was starting to wonder. Was Liz just caught up with taking photos or what? Looking back at my friends in the congregation, I caught their concerned smiles and whispers. After another fifteen minutes, the whispers of concern were getting louder. Was Liz going to leave me at the altar?

Just as I was about ready to go home, the classical musicians we'd hired to play for our guests' arrival started to play 'I'm Kissing You' and the doors to One Mayfair swung open to let the bridesmaids in, followed by Liz and her dad. The image of Liz coming towards me up the aisle will live with me forever. She looked so beautiful. She took my breath away and all my worries disappeared.

She took my hand and there was a moment when time seemed to stand still as we simply stood and looked at each other, trying hard not to cry. We'd made it. We were really there at last, getting married.

The registrar welcomed our guests. We'd chosen traditional vows, not knowing how we could encapsulate the love we had for each other in mere words. Where could we start? There was plenty of opportunity to express ourselves in the rest of the day.

We didn't have readings as such, but Liz's mum had written a poem that had us all in stitches, referring as it did to the time I called Liz's parents to tell them that I loved their daughter. A lovely call to get, you'd imagine, except that I made it at two o'clock in the morning. From a night-club.

'I'm here with your Liz,' I'd told Liz's mum when she picked up. 'I just want to tell you I love her.'

'Tell him to bugger off, it's the middle of the night,' came Liz's dad's voice in the background.

We signed the register, then Liz and I walked back up the aisle together to the sound of the gospel choir singing 'Oh Happy Day', while our guests threw ivory rose petals that whirled around us like flurries of snow. Once our guests were seated for dinner, we made our first big entrance as Mr and Mrs Fletcher, showered by confetti from confetti canons, dancing our way to the top table while Frank Wilson's 'Do I Love You (Indeed I Do)' played. It's a Northern soul classic from my childhood that's since become our song.

Music played a massive part in our wedding day. Many grooms are happy to let the bride get on with the business of wedding planning, but Liz and I had planned our big day as a team. We'd organised every aspect of the day together, from the colour scheme to the menu to the entertainment. There was, however, one part of the day that Liz was not expecting.

Years before Liz and I even got engaged, I went to an event at the Harewood Estate, where the replica *Emmerdale* village was built. It was a great night. One of the things I enjoyed the most was the 'singing waiters', who burst into snatches of well-known opera arias in between serving the dinner. I was so taken by them that I asked for the group's contact details. I had an idea that one day I'd be organising an event where I could use their talent. They were called the Three Waiters.

Over the years, I saw the Three Waiters at several other events and liked them more each time. So of course I knew we had to have them at our wedding.

It was magical. There's something about opera singing that gets me right in the heart. Seeing Liz's face as the waiters put down their plates and burst into the 'Toreador Song' from Bizet's *Carmen* was one of the best moments of the day.

After dinner came the speeches. The first person to stand up and speak was Liz's dad. You could hear how proud he was of his daughter in every word he said.

Liz's brother Daniel made a speech too. Daniel's a surgeon. He had us all rolling with laughter. Then came Brayden and Dean. Brayden and Dean were my two best men. It just so happened that they were also my brothers. Everyone managed to keep it clean.

Dean loves to sing – he and Liz are always talking about doing a duet together – so the minute he stood up with his notes, Liz said, 'Give us a song!' and Dean started to sing 'Amazing Grace'. As his voice filled the room, I couldn't stop crying.

I know it was hard for Dean to stand up in front of a crowd that day, but he made a beautiful speech, which had both me and Liz bawling our eyes out. He took the mickey a bit, as brothers do, but he also spoke with real love about me and his new sister-in-law. Dean's speech was selfless and full of courage. I was so proud of him that day.

It was amazing to be able to share our big day with the people who meant so much to us. My dad had been right to steer us away from a small wedding or an elopement. It meant the world to be able to stand up and confirm our love for one another in front of all the people who had supported us along the way.

As I stood to give my own speech, I looked around that old church and saw people from every part of our lives: family, childhood friends, acting friends, friends who had flown in from all over the world. I saw so many happy faces and knew that we'd been right to bring them all together in one place. I opened my speech by telling them, 'You'll meet a friend for life here tonight.' I'm sure that many of them did.

Our wedding was not just about us and our feelings for one another. It was a celebration of everyone we loved.

LIZ

After all that emotion, it was time to hit the dance floor. Kelvin and I chose the Carpenters' classic 'We've Only Just Begun' for our first dance as newlyweds. We loved that song and it had such significant lyrics for us. Though Kelvin and I had been together for years and we'd known each other forever, in a way we were only just beginning with a whole new phase in our relationship.

It wasn't long before our guests were on their feet too. We'd hired a cover band from Yorkshire to play us into the small hours. Then it was my turn to surprise Kelvin. I kicked off my shoes and took to the stage myself. Together with Michelle, I belted out Creedence Clearwater Revival's 'Proud Mary'. It was the song we always sang together, and we blew the roof off. After that, we partied on into the night. Nobody wanted to go home.

There was only one part of our wedding-day plan that got chopped on the day. A few weeks before the wedding, I'd joined Kelvin on *Emmerdale*'s outdoor set and we'd roped in one of the cameramen to help us make a jokey video to show our guests at the reception. It featured a sort of dream sequence in which Kelvin and I acted out vignettes in which we played each other's 'perfect' partners. Kelvin's idea of a 'perfect' woman was one who could handle farm machinery, so he arranged for me to have a go on one of the *Emmerdale* tractors. When we watched the video back – just the two of us – we decided it wasn't as funny as we'd hoped and it shouldn't see the light of day, but I did have a really good time driving that tractor …

*   *   *

The day after the wedding we stayed in London to make the most of having so many friends and family members in the same place. Because it was getting near Christmas, the

Winter Wonderland funfair had come to Hyde Park. An afternoon there was the perfect way to keep the celebrations going.

The following day, Kelvin and I flew to Rome for our honeymoon. After all the craziness of getting married, it was great to be just the two of us again. Our honeymoon wasn't long – we only had three days – but it was really relaxing. We hired a scooter and went zapping around the city like locals. We saw the sights and visited loads of traditional Italian restaurants, making sure we tasted plenty of wine while we were at it.

Arriving back home as Mr and Mrs Fletcher was really special. We'd known for a long time – perhaps even from our first date – that we were destined to be together, but now we'd made it official. We were partners in crime, ready for the next chapter. Being married took our love to a whole new level. What we didn't yet know was that we were already on our way to being a family of three!

CHAPTER FIFTEEN

END OF AN ERA

KELVIN

What was our new life as the Fletcher family going to look like?

One January morning after Liz and I got married, I arrived at the *Emmerdale* set early and took a moment to just look at the hills in which the village was nestled before I headed on into make-up. It was so beautiful out there in the countryside. So tranquil. As I looked out over the wintry fields, something about the view seemed to speak to me and I imagined a small child – our child – laughing as she fed handfuls of grass to a horse.

What would it be like to raise a family on a farm? I asked myself as I stood there taking in the sights, the smells and the sounds of the landscape I had come to know so well over my years spent playing Andy Sugden. Could farming be the life for me?

No. I shook my head. I may have played a farmer for most of my adult life but that didn't mean I knew anything about being one. It was just a ridiculous dream ... I kept coming back to it, though.

The previous year, 2015, had been a big one for me and Liz. It wasn't just that we'd decided to get married. We'd been preparing for other changes too.

I'd been on *Emmerdale* since I was a child. It was a hugely important part of my life. The *Emmerdale* cast and crew were like a second family to me. But just as I'd had to fly the comfort of my childhood home to be able to eventually create my own family with Liz, I'd begun to think that perhaps it was time for me to move on from *Emmerdale* too.

It wasn't that I was unhappy on the show. I was the happiest I had ever been. *Emmerdale* itself was the best it had ever been too, winning awards left, right and centre. But after twenty years as Andy Sugden, there were other things I wanted to do. I wanted to do film, I wanted to do comedy, I wanted to be on the stage ... but my *Emmerdale* contract prevented me from doing anything else.

At an ITV dinner, I chatted to my friend Steve November who had left *Emmerdale*'s production team three years previously. Listening to him talking about the work he'd done since then was inspiring. He'd moved into drama and was commissioning shows like *Beowulf* and *Broadchurch*. That was the kind of stuff that interested me now. I would have loved to do a show like *Broadchurch*.

I told Steve as much. He seemed surprised. 'We just assumed you were happy doing *Emmerdale*,' he said.

I *was* happy doing *Emmerdale*, but I knew I also wanted to do so much more. Talking to Steve was a wake-up call. I thought, *Maybe everyone looks at me in the same way and thinks I don't want to play anyone but Andy Sugden. They think I don't have any other ambitions.*

I couldn't sit on the fence any longer. I had to believe in myself. I had to show people like Steve that I meant business. That meant leaving Andy Sugden behind.

I spoke to Liz first. Whatever decision I made would affect her too, so she had to be a part of it.

If I was nervous about telling Liz I wanted to leave *Emmerdale*, I needn't have worried. Her reaction was exactly what I hoped it would be.

'Kelvin,' she said, 'you've got to do what you've got to do.'

Liz understood what it was like to be in a situation that didn't feel quite right. She immediately saw the parallels between my *Emmerdale* dilemma and the moment she decided she wanted to leave her job in fashion to go to drama school. That fashion buying job was one many people would have jumped at, but it wasn't right for her.

Just as I had encouraged Liz to take a leap of faith back then, she was ready to do the same for me.

She reassured me, 'There's nothing more attractive than someone following their dream.'

Liz made it clear she would be behind me all the way. It's what being a good partner – and a good friend – is about: wanting the person you love to be as happy and fulfilled as they can be regardless of how their decision might affect you personally.

With Liz backing me up, I felt ready for anything. Which isn't to say that I didn't face some questions about the viability and sense of my future plans. Dad in particular was much less sure I should let *Emmerdale* go. Like Liz, he wanted me to get the most out of life, but he also worried about the financial implications of leaving a recurring role in one of Britain's biggest serial dramas. There weren't many jobs an actor could get that offered that kind of security.

Dad didn't have anything growing up and he'd grafted for everything he had now. He understood more than anyone the importance of financial independence because he knew what it was like to have to count every penny. He thought that I already had a dream career.

It was hard to hear Dad say he thought I was making the wrong decision, but I knew he was only looking out for me. Gradually, I convinced him that there was more out there for me. I also reminded him that I'd been very lucky

Liz: Backstage with Michelle after a performance of *Joseph and the Amazing Technicolor Dreamcoat* at the Palace Theatre Manchester.

Liz: Blowing out the candles at my tenth birthday party with Matt and Kelvin by my side.

Kelvin: One of the first dates we went on. Liz is looking good in her infamous Mr T jewellery . . .

Liz: A very proud moment graduating from university with my degree in fashion buying.

Kelvin: Batman and Robin at the 2005 Challenge Cup Final in Cardiff. A typical weekend away wlth Crowther.

Kelvin: Enjoying the slopes on a surprise Christmas skiing trip for New Year, Val Thorens, France.

Kelvin: Our trip to Paris, where Liz was half expecting me to propose . . .

Kelvin: . . . And the actual proposal! Sunset on the Menai Strait, Anglesey, 28 November 2014.

Kelvin: Forty-eight cows, one pig and Andy Sugden . . . The stag do heads to Tenerife!

Liz: Way too many Lizzes in one picture! Celebrating my hen weekend in Marbs with my closest friends.

Liz: Posing with my bridesmaids – we had no idea we were running so late for the wedding!

Kelvin: Looking on as Liz's mum, Mary, reads her poem during the wedding ceremony.

Liz: Past my due date with my first pregnancy.

Kelvin: New parents to Marnie Molly Fletcher.

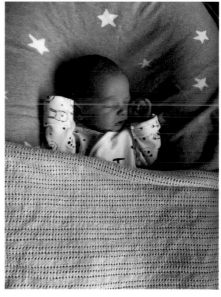

Liz: First night home with Milo.

Kelvin: Our first ever agricultural show. Marnie is showing her Cotswold ewe lamb in Ashbourne's Young Handlers category and coming third!

Kelvin: Like father, like son. Milo helping Daddy fix something.

Kelvin: Market day isn't market day without ice creams!

Kelvin: Exactly what I used to imagine living in the country would look like. A dream come true.

Kelvin: Little Milo helping to feed the sheep.

Kelvin: Although not your typical farm dog, Ginger has adapted very well to country living.

Kelvin: The future farmer. What a great shepherdess she'll make.

Liz: Flying solo with the kids and Ginger, off on one of our many adventures around the farm.

so far, having so many years in one role. Who was to say that luck would last even if I didn't give *Emmerdale* up? Who was I to think I'd have another twenty years as Andy Sugden? And so what if I did fail? I had him and Mum. I had Liz. I knew they would be there for me no matter what. In my eyes, I wasn't risking as much as Dad seemed to think.

Success to me was never just about the figures in my bank account, it was about having the courage to live life to its fullest. I didn't want to get to seventy and look back with regret, wondering what I might have done.

Dad eventually agreed. 'You've got to follow your heart,' he said. 'Just be careful!'

*   *   *

I was careful. Having made my decision, I sat on it for another six months. During that time, I did my research. I kept a close eye on the sort of shows that were being commissioned and asked myself honestly what parts I thought I could play. I realised that as far as acting was concerned, I was reaching an 'awkward age'. I was no longer young enough to play the classic hero, but at the same time I wasn't old enough to play, for example, a convincing father of four.

About this time, fate sent a curveball. Since the age of eight, I'd been represented by agent Peter West. Peter was a great man. He was lovely and warm, very 'showbiz' and larger than life. He couldn't have been more different from the people I'd grown up around in Oldham. When you think of the archetypal actor's agent, that was Peter. He was a lovely soul and a fixture at every party. I could really open up to him. I felt as though I could be myself around Peter because that camp showbiz style was in my DNA too.

Peter had been ill for a while, but he was a very proud man and he didn't want to talk about his illness. So when he died it came as a shock. We'd had no idea how unwell he really was.

The loss was two-fold. I felt I'd lost a true friend, someone who had always been there for me, and of course I'd lost my agent. Suddenly, I was making big plans for my career without the experienced guidance I'd been used to.

It was a tricky moment. I knew I needed another agent, but what other agent would have me? I was a well-established actor in a longstanding role, but if I was going to be true to myself, I would have to tell any new agent what I was planning; that if they wanted to take me on, they might not be making any money out of me for a while.

Liz suggested I arrange a meeting with Olly Thomson. I liked Olly instantly. But how would he feel when he heard I was planning to cut loose from *Emmerdale*?

To my surprise, Olly listened carefully and suggested, 'Let's make a plan.'

That's when I knew he was the right agent for me. He wasn't just about making money. He knew that, as an actor, I wasn't just a cash cow. He'd trained as an actor himself and understood why I'd been feeling unfulfilled.

At this stage, I had not yet made my decision to leave *Emmerdale* official. I hadn't even hinted that I wanted to leave. Olly suggested that I do one more year on the show, to enable me to prepare myself for what might be a bumpy ride ahead.

I was nervous about making my plans public, but when it came to telling the production team, they were surprisingly understanding. At first it was suggested that I take a sabbatical, but that wasn't what I wanted. I had too much respect for the show to keep it as a backstop if things didn't work out elsewhere. I knew that I had to make a proper break, so we agreed that Andy Sugden would be written out of the show.

Once the truth was out there, I felt immediate relief, but I took a little longer to tell the people I worked with most frequently that I was going. I didn't want the thought of my departure to be hanging over all our interactions for too long, so I kept it under my hat until the last two to three months.

There were about half a dozen people that I wanted to tell first, before the rumours started. I didn't want anyone to hear via a leak. I wanted to honour the working relationships and friendships I had on set. I was especially worried about telling Danny Miller, who played Aaron Dingle. He was always flying the flag for *Emmerdale*. His reaction was simple: 'Kelvin, I'm gutted.' People were sad to see me go and when they expressed as much, it made it hard not to wobble. But if you don't take a chance, you can't evolve, and I needed to be doing something different. Before long, everyone on set knew I was leaving and there was no turning back.

* * *

There are many actors out there who wouldn't do a soap. They think it's beneath them. That's reflected in the wider industry, such as at an awards ceremony where 'best actor' and 'best soap actor' are two different categories. Why is that? It's no easier to give a convincing performance in a soap than it is to walk on the stage at the National Theatre.

Judging soaps and their viewers is naive and narrow-minded. A soap attracts up to ten million viewers every night and the connection with the audience is profound. I realised just how profound when I started to get fan letters.

At the *Emmerdale* studio we each had a pigeonhole, where the producers would leave any mail we or our characters received. Steve Halliwell, who played Zak Dingle, got the most. His pigeonhole was constantly full. I was pleased to get a fair bit myself. It wasn't just that the letters were flattering, they were also really interesting. Because my character had always been a tearaway, I appealed to a really diverse group of people. At one end of the scale, I connected with kids who saw themselves in young Andy. At the other end were the grandmothers who told me I reminded them of their grandsons.

'I know it's only pretend,' they would write. 'But I can't tell you what it means to me ...'

To my surprise, I also received letters from a number of people serving time in prison who wanted to give me – or rather Andy – advice before my life went off the rails for good.

'Don't end up doing a ten-stretch like me ...'

'You've got a chance to do things differently if you change right now.'

So many people wrote to tell me that, watching Andy go through life, they suddenly felt seen and heard themselves. Their letters made me understand the responsibility and privilege of playing a long-term character on a serial drama. Andy Sugden was someone these viewers felt they knew; he was almost a friend. In playing that part, I was touching people's lives and, without knowing it, bringing some comfort. I was making a connection with people I would never meet. The letters taught me that you should never judge people. The ones I received from prison inmates were the most humbling. It must have taken courage for some of those people to be so honest about their own experiences.

As Andy's storylines changed, so did my correspondents, but they always reminded me of the importance of having the courage to put real emotion into my performance and to approach each new plot twist with authenticity. They'd taught me that someone out there might see Andy Sugden face his problems and gain the courage to face their own.

My final *Emmerdale* storyline was a big one, in true Andy Sugden style. Framed for a murder he didn't commit by his former lover Chrissie, Andy was awaiting trial in prison when a visit from his barrister gave him the chance to escape. Going back to Home Farm, he confronted Chrissie, who cut herself in an attempt to frame Andy again, leaving Andy no choice but to go on the run. His brother Robert got him a false passport and, after a series of emotional goodbyes by his dead wife Katie's grave, Andy left the village, never to be seen again.

As I filmed those final scenes, I was crying in character – as Andy – but there's no doubt that deep inside a bit of me was crying too.

In many ways, Andy Sugden was the polar opposite to me. I'd never steal. I'd never knowingly cause another person harm … But then I grew up in a very different environment. I come from a strong family, which instilled in me the values of honesty and hard work. If I'd come from a broken home, with no mother and a father like Andy's, who spent time inside, things might have turned out very differently. Playing Andy made me realise that we all have flaws, we all face challenges and we all handle them differently according to the hand we've been dealt. We are all capable of making mistakes and everyone deserves a second chance.

*   *   *

On my last day of filming, which took place in late summer, my bosses came down to the set with champagne and gave a speech.

Though I had been preparing for this moment for a long time, I still couldn't quite believe it was here at last. Liz came to the set for my leaving do. By this time she was heavily pregnant.

Looking back, I wonder if leaving *Emmerdale* after so long would have been harder had it not been for the fact that so much was going on in my personal life. It was difficult to mourn the loss of that one part of my life when I had so much to look forward to elsewhere.

But it was the end of a chapter. A brilliant twenty-year, 1,100-episode-long chapter. I'll be forever grateful to

Emmerdale and ITV. It was almost as if for two decades, I'd got to live two lives – my own and Andy's. Thanks to Andy, I'd experienced things that I never would have done in my own world. At the heart of everything, Andy Sugden was a good guy and I was proud to have been able to bring his story to life, but I was done being a fictional farmer. It was time for the next stage.

CHAPTER SIXTEEN

AND BABY MAKES THREE

LIZ

By the time it came for Kelvin to leave *Emmerdale*, I was massively pregnant and we were looking forward to our baby's arrival any day. But my due date came and went and there was still no sign that Baby Fletcher was ready to make an appearance. I'd heard from just about everyone that first babies were sometimes late, but it seemed that our baby was hoping to set a record.

Finally, at long last, my contractions started. They came on in the middle of the night. I lay there in bed, completely awake, while Kelvin slept peacefully beside me. The second he opened his eyes in the morning, I was onto him. By now, my contractions were so strong I honestly felt like I was going to die.

'Kelvin,' I said, 'we've got to go.'

I didn't even let him have a coffee.

We thought we were well prepared for this moment. Over the past few months, Kelvin and I had attended a series of hypnobirthing lessons where we learned techniques that we hoped would enable me to give birth without needing drugs or any other intervention. The course was designed to reset our perceptions of birth. The instructor reminded us that giving birth is not a problem but a natural process. 'Like we breathe, we reproduce,' was a hypnobirthing mantra. It seemed to make sense. I was designed to give birth. Women had been giving birth for millennia. It was a perfectly natural thing to do.

We thought we had it nailed. We planned that I would give birth in the luxurious new birthing suite at Oldham Hospital while Kelvin read poems to help me stay calm. Even as we got into the car to drive to the hospital, we thought we knew how everything would pan out.

Kelvin is good in stressful situations. He knows how to keep calm and make everyone around him feel safe. On the way to the hospital, he did his best to channel that energy, though every time we went over a speed bump I wanted to scream in pain.

We got to the hospital and I was about to get out of the car when I spotted someone I was sure I knew walking across the car park. I *did* know her. It was Sally Johnson, the friend whose mother's car I'd bunny-hopped into the back of a white van back when we were thirteen.

'Hide!' I told Kelvin as I lay down flat on the back seat so that my old school pal couldn't see me. Sally and I had drifted apart since we were kids and while I was sure she was a very nice person, she was not someone I wanted to see on my way to give birth. What were we going to do? Make small talk between my contractions? Hiding seemed the only option.

'Kelvin, get down!' I yelled.

Kelvin ducked behind the steering wheel so that Sally couldn't see him either.

We were like that for just a few seconds before I got another contraction and howled and Kelvin said, 'Liz, this is ridiculous! Come on. Don't worry about Sally – we can't sit here all day.'

I might have sat there all day, but Kelvin hauled me out of the car and we raced for the doors of the birthing suite. As we waited for the lift, a patient pointed at Kelvin and shouted '*Emmerdale*!' Kelvin responded graciously, but we couldn't wait to be somewhere private. Our hypnobirthing course had not included information on what to do if people keep wanting to talk to your famous actor husband while you're in the final stages of labour.

It was a big relief to get into our room. It was like a spa in there, with soft lights and soft music playing in the background. In these surroundings, we were sure that I would be able to give birth quickly and calmly. When it came to it we didn't follow any of the hypnobirthing rules apart from trying to have a calm mindset and playing calming music. We tried to focus on the breathing, but two hours later there was still no sign of the baby.

I'd been given gas and air but as the contractions got closer together I knew I needed something stronger. I told Kelvin, 'They've got to give me an epidural.'

When we'd been writing our birth plan, rather naively I'd told Kelvin that I did not want an epidural and under no circumstances was he to allow me to have one. If I said I wanted one, he was to stop me, but this was all before I was in labour, when it was still a distant dream. But Kelvin being Kelvin and always wanting to do right by me thought that it was his job to stop me getting one, telling the midwife to ignore me.

'She doesn't want one,' he said.

I was bent double, crossing my arms around my body, having another contraction.

'I want an epidural!' I insisted.

'You don't need one,' said Kelvin, from his chair.

I assured him I bloody well did.

While all this was going on, the nursing team were busy taking readings. They were calm and efficient, but I could tell that something wasn't quite right. My instincts were spot on. The baby's heartrate was dropping. It wasn't a good sign.

In a moment, we were out of the spa-like birthing suite and on our way to the hospital's main labour ward. It could not have been more different. There was no fairy music here. Just clinical efficiency as I was hooked up to monitors for everything imaginable.

When that was done, I was told, 'Liz, you've got an hour. If you can't push any more, we'll have to pull.'

But having at last had the epidural I'd begged for, I couldn't feel anything and pushing seemed impossible. The baby's heartrate continued to drop. The doctor stepped in with the forceps, just as promised. Kelvin stayed at the top of my bed by my head, too squeamish to watch the birth.

'I'm going to do 10 per cent of the work,' the doctor told me.

Ten per cent? The first pull dragged me halfway down the bed! It seemed hopeless, but three pulls later, our baby was out.

The doctor called Kelvin over to the bed and handed our baby straight to him.

'You can tell Liz what the baby is,' he said.

We'd decided not to learn the sex of our baby during the antenatal scans, looking forward to exactly this moment.

Kelvin gently unfolded the blanket swaddling our newborn and told me, with tears in his eyes, 'He's a boy, Liz. He's a boy!'

Everyone else in the room looked confused.

'Have another look,' the doctor suggested.

Our baby was not a boy. Overwhelmed by the emotion of the moment, Kelvin had mistaken the umbilical cord for a penis. Later he admitted, 'I did think, *Bloody hell, that thing's never-ending!*'

With the true sex of our baby established, we were left alone in the labour room for a while. It was a moment of true calm and stillness, and we savoured every second of it, knowing that we would never have this moment again. It was indescribably profound and wonderful to be meeting our daughter for the very first time.

But unfortunately, not long after, reality hit. While Kelvin and I were basking in the bliss of becoming new parents, a doctor returned to make sure that I was OK. After such a long labour, which had climaxed in a difficult birth, I already knew I would need stitches. But it was more complicated than that. I still couldn't feel a thing below the waist but apparently I was bleeding heavily. The doctor instructed me to pass the baby to Kelvin. I was going straight to surgery.

I was in the operating theatre for the best part of three hours that day, while a surgical team worked hard to stop me from haemorrhaging. I'd later learn that they had to give me a massive blood transfusion. When I woke up from the anaesthetic, I was exhausted and disorientated. I couldn't stop being sick. And I was worried. I had no idea where Kelvin and our baby were.

Still in a daze, I was taken from the operating theatre to a high-dependency ward where I would have to stay for the next five days. When Kelvin was allowed to bring our daughter in to see me, I burst into tears of relief.

Those five days in hospital were a difficult but ultimately happy period. Kelvin was allowed to stay with me and that's where we got to be parents for the very first time.

In the months running up to our daughter being born, we'd tried out loads of different names. We hadn't dialled in names exactly – we wanted to wait for that 'Eureka!' moment when we met our baby for the first time – but one of our favourite girls' names was Marnie. Marnie Molly. My nana was called Molly and Marnie is Molly in Irish.

As we got to know our baby girl, we were sure that Marnie would suit her perfectly.

Despite the drama surrounding her birth, Marnie was a very easy newborn – a dream baby. She barely cried. She just seemed to be happy to be out in the world. Still, after those five days in hospital, when we were told we could take her home, we couldn't help but have a moment of panic.

'They're just going to let us get on with it?' I asked incredulously. 'Are we ready to be parents?' Deep down I thought, *OMG, is that it? You literally get sent home with a baby?* I am not sure that I was expecting an operating manual, but I would have been pleased to get one. It was all so surreal, but we had no choice. The hospital needed my bed back!

* * *

Back at home, we tried to settle into a rhythm. It wasn't easy. As any first-time parent will recognise, those early days were a blur. Everybody wanted to visit and, not quite understanding the kind of support we needed, we let everybody and anybody come to the house. It was intense. It was a lot to be new parents and at the same time to feel we had to throw open our doors to everyone, especially given what I had been through. I mean, I'd almost died. Under what other circumstances would anyone expect someone who'd just had a three-hour surgery and a massive blood transfusion to even feel well enough to get out of bed? The very last thing I needed was a string of visitors sitting there looking at me in my most personal moments.

I didn't know how to take care of myself in the face of so much change. Shortly after we got back from the hospital, I cried for two weeks solid. The crying was uncontrollable and was set off by everything and anything. I remember sniffing the little hat that Marnie had worn as a newborn and bursting into tears, while Kelvin tried desperately to comfort me. He was so worried about me.

'What can I do, Liz?' he asked. 'Let me know how I can help.'

I couldn't even begin to articulate what I was feeling. It was the strangest mixture of anxiety about being a good mum and a love so overwhelming I could hardly bear it.

Before we'd left the hospital, a member of the team there had told me all about the counselling they offered new mums – particularly mothers who have experienced traumatic births. I didn't think I needed it and didn't take them up on it. In retrospect, I can see that I was still on a high at that point, and the reality didn't hit me until we were safely back at home. It's such an extraordinarily overwhelming experience, with so much emotion mixed in. I just hadn't realised how naive you are with your first child. How, even though you can read all the books, go to all the classes and prepare everything, you're still unprepared for that moment when you hold your child for the first time. The changes whoosh in all at once – not just to your body, hormones and emotions, but to everything else as well, as you quickly realise that life will never be the same again. I guess a lot of first-time mums get that, and I was no different. I just pushed on through the tears and the worry and, thankfully, after a while I did start to feel better. I understand now that what I experienced was the 'baby blues'.

My hormones were all over the place and my having had a general anaesthetic must have been a factor too.

I got through it, but I know now that I was lucky. Since Marnie was born I've learned more about other women's post-natal experiences and wish I'd taken the hospital up on their offer of counselling so that I might have understood at the time that what I was experiencing was absolutely normal and nothing to be ashamed of. I would advise anyone who finds themselves in the same situation to just reach out. You're not alone.

Fortunately, having left *Emmerdale*, Kelvin was able to take some proper paternity leave. He was so hands-on, always willing to change a nappy. I couldn't have hoped for a better partner at such an important time in our lives. Soon Marnie gave us her very first smile and the difficulties of those early weeks faded into the distance.

CHAPTER SEVENTEEN

BEST FRIENDS FOREVER

KELVIN

It was bliss to be able to take three-months' paternity leave to be with Liz and Marnie. I enjoyed every moment, from the night feeds to the nappies, as I watched our daughter blossom and grow. I knew I was lucky to be able to take so much time to just enjoy being a first-time dad – statutory paternity leave in the UK is a maximum of two weeks – but as time rolls around so quickly I also knew I had to put myself back out there and get a new job. My slow exit from *Emmerdale* had given us time to save up and prepare for Marnie's arrival, but our savings wouldn't last forever. I had to start bringing money in again.

I had been looking forward to spreading my wings and diversifying, but over the months that followed my telling Olly I was ready to work again, it felt as though I couldn't get a break. I hadn't expected to walk straight into the next Bond film, but I had expected someone to see a way to use my potential. I felt like I was constantly trying but never getting anywhere.

But when I did get work, that was hard in another way. Those three months of paternity leave with Marnie had been amazing and I hated to be away from her and Liz for more than a couple of days. Stuck in a hotel room on location, I felt like I was on a different path. In order to support my family, I had to sacrifice my family life. I felt really lonely but what could I do?

There was one person that I felt I could always turn to for advice. That was my best mate, Crowther.

Crowther actually started out as a friend of my dad's. He was about five years older than me but as soon as I met him, I knew we would be great mates too.

We couldn't have been more different. To begin with, we were physical opposites. Crowther was a rugby player. He had the potential to be world class but lost an eye when he was young. While I was driven, he was laid-back. I wanted to live and work all around the world. Crowther was happiest in Oldham.

Despite our differences, we became the best of friends. He was always ready for a laugh. We'd play hilarious practical jokes on each other. When I had days off and Liz was working, Crowther was the one I always called. We'd start the day at the gym, then go to a local carvery – £5 for all you could eat. We'd go on holiday together. Sometimes, at the end of one of those carvery days, we'd just drive to the airport and jump on the next flight. It didn't matter where. Liz got used to me calling from Manchester, asking if she wanted to join us on a last-minute trip to Spain.

'Some of us have got to go to work,' would be her dry reply.

I once took Crowther with me as my wingman to a TV awards show. Somehow, while I was elsewhere, he got chatting to Dale Winton, the *National Lottery* host. When I found them together, Dale was roaring at one of Crowther's jokes.

'Your mate is the funniest man I've ever met,' Dale told me, eyes glittering with tears of laughter.

Crowther was always completely unfazed by spending time in the company of celebrities, treating them just as he would anybody he met in the pub.

Another time, I got tickets for the RFL Challenge Cup Final, a big Rugby League match in Cardiff. For some reason, we decided that we would go to the match in fancy dress. Crowther wore a Batman costume, while I got to be Robin. Ahead of the match, we went to a pub to have a few drinks. We weren't close to stadium and didn't notice time passing until we'd almost missed the start of the match. Dashing towards the town centre with our half-finished beers, we turned a corner and came face to face with a police riot van.

I thought we were in trouble. On match days, there are strict rules about drinking in the street in Cardiff. There's an open alcohol ban and there we were, carrying open cans. My heart sank as the side door to the van slowly opened and one of the officers inside beckoned us over.

'Come here, lads.'

This was it. I was sure we were going to be arrested. I immediately started thinking of what the producers on *Emmerdale* would say if it got splashed all over the press. Instead, one of the police officers asked, 'Aren't you supposed to be at the game?' Then he offered us a lift, dropping us right outside the stadium, to the amusement of everyone who'd ever wanted to see Batman and Robin jump out of a riot van.

It got even better when we were inside the stadium. I'd been invited to go upstairs to the executive boxes where the chairman of the RFL was holding a black-tie dinner. Everyone but us was suited and booted.

'What are we going to do?' I asked Crowther, when I realised how far we'd strayed from the dress code.

'What would Batman do?'

Crowther pulled his mask down and carried on in.

As I tried to follow him through the crowd of amused Rugby League dignitaries, I was cornered by our host.

'Where's Batman?' he asked.

'Over there by the free bar,' I said simply.

The laughter told me we'd got away with it.

Crowther was always ready for a laugh. At the same time, he was surprisingly full of wisdom, and I often turned to him for advice. He had a different way of doing things, but he always seemed to understand the dilemmas I faced and have a useful perspective. If he thought I was about to make a mistake he would say so, but always in a kind and gentle way.

'Don't miss any chances, Kelvin,' he'd tell me. We spoke every day. He was an incredibly important part of my life.

Around the time I left *Emmerdale*, Crowther was going through a rough patch. He'd recently had a son, who was the apple of his eye. Unfortunately, Crowther's relationship with his son's mother had hit the rocks and he found himself having to go to court in order to get visitation rights. It was a long and painful process that laid Crowther really low. We were all so pleased for him when we heard that he had been granted the right to see his son again.

With that good news under his belt, Crowther had started to make plans for the future and seemed to be getting back to his old self. When Marnie was born, Crowther was really happy for me and Liz. He was kind and supportive, giving us lots of advice on how to cope with a newborn. We found it hilarious in many ways –

that the Crowther we'd always known as such a reprobate could have turned into a doting dad and a fount of useful parenting information.

As new dads, Crowther and I would sit in the pub, imagining our children growing up side by side and having a friendship every bit as strong as ours. There was so much to look forward to. Which was why it was such a shock when partway through October 2016 I was woken at two in the morning by a phone call from another friend of ours, Andy, Crowther's brother-in-law.

'Steve's in hospital,' he told me. 'He's had a heart attack.'

*   *   *

I got to Crowther's bedside as soon as I could. I couldn't believe he was in hospital. The heart attack had come out of the blue, as far as all his friends were concerned. Crowther was one of the fittest people I knew. He was always in the gym. It was terrible to see him in that hospital bed looking so ill.

'Mate, what have you done to yourself?' I asked him.

The cardiologist fitted Crowther with a stent to keep blood flowing to his heart and sent him home a few days later with strict instructions to improve his lifestyle. That included minimising the stress in his life. To my mind, there's no doubt that the stress of the court case to get custody of his son contributed to what happened.

We were all behind Crowther, ready to help him do whatever it took to get back on his feet again. 'You've got to rest up,' I told him. 'This is your wake-up call. You've got to take better care of yourself so you can be here for your son.'

Crowther assured me that he wasn't going to mess around. He knew he'd been lucky. The minute he was back at home, he set about doing everything the doctors suggested. He changed his diet and lost weight. He was back at the gym. He was going to make the most of his second chance.

In January, Crowther joined me and a few friends to celebrate my birthday at a restaurant in Oldham. It was a low-key night – not like the ones we'd had back in the day before Crowther got ill – but it was still great. Crowther was on good form. He looked amazing. He told me he felt lucky to be alive.

I remember him laughing about his weekly NHS physio sessions at the hospital, where he joined five other heart patients for a workout. At thirty-nine, Crowther was the youngest person there by forty years!

'They're all stronger than me, mate,' he joked.

Still, he told me he was going to stick at it. He would do whatever it took to make sure he was around for his boy. No one was prouder of Crowther than I was that night.

Back at our respective homes after the meal, Crowther and I sent a few texts about how much fun we'd had, and how much we were looking forward to next time. He didn't respond to my last message. I guessed he must have fallen asleep.

*   *   *

I think I knew the minute I heard my phone ring that it was bad news. How could it be anything else at two in the morning?

I answered the call. It was Crowther's brother-in-law Andy again. Crowther had had another heart attack. This time he hadn't survived.

The news hit me like a brick in the face. I got out of bed and went downstairs, feeling like I had just walked away from a crash. I got in my car to drive to the hospital to see him. My route took me right past Crowther's house. A police car was parked outside and the front door was open. Feeling an overwhelming need to be close to Crowther as I remembered him, in a place where we'd had so many good times, I stopped. A policeman let me inside.

I'd been in that house so often it was like my second home. Since splitting up with his son's mother, Crowther had lived alone. He wasn't exactly house-proud, and I'd often helped him tidy up before his son came over. Crowther's dog was waiting in the kitchen. I stooped to give him a quick scratch behind the ears but he wasn't interested. He was agitated, wondering what had happened, watching anxiously for his master's return. Then I went upstairs to the bedroom. My heart broke when I saw his familiar glasses on the bedside table.

I carried on to the hospital where Crowther's dad and Andy were waiting in a private room. Seeing Crowther in that hospital bed, so still and silent, was surreal. How could my friend, who was always so much larger than life, who filled every room with laughter, really be dead?

Back at home, I fell to pieces. Liz did her best to comfort me. She knew that Crowther was more than just a friend to me. She used to joke about him being my 'second wife'. He was a soul mate.

I couldn't believe it. The previous evening we'd been laughing and joking, ready to embark on the next chapter in our lives. We'd been planning to take our kids away on their first holiday together. Now Crowther was gone. There would be no third chance.

* * *

Over the following days, I realised that I'd been lucky until I lost Crowther – I'd never before experienced such a devastating kind of loss. Now I didn't know how I would ever get over it.

Crowther's funeral took place a couple of weeks after his death. I was a pall-bearer and gave a eulogy. The church was packed – a testament to just how well loved he was by everyone who knew him. His loss would be felt by a great many people in Oldham. He was mourned by the whole town and beyond. He was such a pure, kind-hearted man.

As we stood in that church, I felt empty. I was surrounded by great people. I had Liz and Marnie in my life. But the loss of Crowther had knocked me for six. It had shaken me, Crowther dying so young. When I looked at his son, I was overwhelmed with sadness, and it scared me as the father of a small child myself. What if something happened that meant I couldn't be around for Marnie?

Crowther's death made no sense. He was a good man. He'd been trying to turn his life around. It wasn't fair. As I thought that, I could almost hear Crowther laugh and tell me, 'Life isn't fair, mate. None of us knows how long we've got. You've got to make the most of it.'

Was I making the most of my own life? Who was I? What did I want? I didn't seem to know any more.

Crowther's death marked the start of a very dark period for me. In the weeks that followed his funeral, I went deep inside and examined every aspect of my existence. I knew that family – Liz and Marnie – was what mattered most to me. They were my highest priority. Being able to provide for them was important, of course, but I also wanted us to build a life that was fulfilling on other levels too. How could Liz and I do that?

Liz was endlessly patient and wise as I tried to work out where to go from here. But the irony was that the one person who I felt could really have helped me work through the dark days without Crowther *was* Crowther. He would have known exactly what to do.

One afternoon, I went for a walk alone in the Peak District and tried to connect with the feelings inside me and find some guidance in the solitude. The countryside was so peaceful, I thought I'd be able to hear myself think there and answer my own questions honestly. Prior to Crowther's death, I'd been running myself ragged, chasing down my dream acting career. Had I been running in the wrong direction?

I looked across a valley to a small farmhouse nestled in the shadow of a hill and wondered who lived there. I'd come just a few miles from Oldham, yet it felt like I'd travelled to a different planet, where the glitter and rush of the television world seemed meaningless. Was this peace and tranquillity what Liz, Marnie and I needed now?

CHAPTER EIGHTEEN
HOLLYWOOD DREAMING

LIZ

Kelvin and I had known each other practically our whole lives, but I had never seen him so down as he was in the weeks and months after Crowther died. He really was at rock bottom. Crowther's untimely death had shocked us all – I, too, had lost a dear friend – but I knew it was ten times worse for Kelvin. Crowther had been so special to him. They were as close as any brothers.

It was hard to see Kelvin suffering. At the same time, I was having a difficult time myself. Suddenly, all the emotional weight Kelvin and I normally shared between us was squarely on my shoulders. And we had Marnie to raise and money to earn. What were we going to do?

At the end of 2017, more than a year out of *Emmerdale*, Kelvin was still feeling in the doldrums career-wise. He was getting the odd acting job but nothing that set the world alight. Meanwhile I was getting a bit of voiceover work that took me down to London. That was great for the bank balance, but it was hard when Kelvin was working, and we had nowhere to turn for childcare but to our parents. We could have carried on ticking along like that indefinitely, but we both knew that something needed to change.

In the depths of the English winter, we came up with a radical solution. Sunshine. We'd both always wanted to live abroad for a bit so we decided that we would do exactly that. Why not? Right then we had nothing to lose.

We settled on Los Angeles – Hollywood – the obvious place for a pair of actors to go. Since Marnie was still too young to go to nursery, it would be easy for the three of us to up sticks for a couple of months and find a new perspective in a very different town. We sold our car to finance the trip, then we booked an Airbnb in West Hollywood and set out to see what California was all about.

From the minute we got there, we loved it. We loved the lifestyle, the endless sunshine, the yoga and juice bars on every corner and the way that everyone was always so upbeat. It was really refreshing. Best of all was the 'can do' attitude that is the American trademark. It felt great.

Being in LA was also strangely liberating in another way. Having been on television for most of his life, Kelvin was used to being recognised wherever he went in the UK. If I had a pound for every time someone asked me if I could take a picture of them with my husband! In the US, there wasn't the same level of recognition and therefore nobody had any expectations of us. We could be whoever we wanted to be. We could leave Andy Sugden far behind.

Crowther's death had taught us that life is short and you never know when your time might be up. Losing Crowther had sent Kelvin to some dark places, but without the experience of working through that pain we might have pottered along forever and never asked ourselves if we were truly on the right path. We would never have gone to LA. It was exactly the tonic we needed. Life seemed full of possibility again.

By the end of our six weeks in Hollywood, we had decided that we were going to go for it: we were going to move out to the States for a couple of years. Kelvin had already got himself an American agent so we set about starting the visa process, which required more paperwork than you can ever imagine, including sending over tons of press clippings, to show that Kelvin was already an established talent in the UK. We also started working out how we would support ourselves while we got going in LA. The fact that Kelvin wasn't as well known in the US was a bonus in that respect. He could get a job in a café in Hollywood and no one would bat an eyelid. Of course, that wasn't an option in the UK. It would have been all over the tabloids. In the US, we would have many more options. It was really exciting to think that we could do anything.

Because of the way the US visa system works, we had to move back to the UK while we were going through the process. And soon after we applied, I fell pregnant again, so we added the new baby to the visa application too. Nothing was going to stop us from following our Hollywood dream.

*   *   *

While we waited for our US visas to come through, which we knew would probably take a minimum of eighteen months, life ticked on and, before we knew it, it was time for baby number two to be born.

My experience of the baby blues and having been so overwhelmed by all the visitors right after Marnie's birth made me determined that things would be very different second time around. I was booked in for a C-section and told our family and friends that, much as we loved them and appreciated that they would want to meet our new addition as soon as they could, we would need our quiet family time. Being able to plan that way helped enormously.

It took us a little while to come up with a name for our newborn son. As in the run-up to Marnie's birth, we'd batted names around, but when he arrived we looked at him and realised that none of the names we'd thought of really fitted. Nor did we have that 'Eureka!' moment, with a new name popping into our heads as if by magic. In the end it was my dad who suggested 'Milo'. We liked it at once and when we tried it out for size, it made perfect sense. Milo was definitely a Milo!

Milo's birth went exactly as we had hoped, but I knew in the run-up that the weeks that followed were going to be tough. The very day after Milo's birth, at six in the morning, Kelvin had to leave for Deptford in East London, to begin rehearsals for a run of *The Wizard of Oz* in Blackpool, where he would be playing the Tin Man.

Of course, Kelvin had no choice but to do the show. We needed the money. But it's fair to say that the experience of being on my own with the children while he was away was hard for me. I don't think I had fully appreciated how full-on it would be to have a toddler and a newborn. I was recovering from a C-section, which meant I could hardly sit up or even stand straight to begin with. I certainly wasn't supposed to do any heavy lifting. Try telling that to a toddler! I think I only managed to get through the first few days because I still had morphine in my system.

But the morphine wore off. After that, life was a blur of night feeds, nappy changes and nursery runs. I barely had time to brush my hair. It was tough to be apart from

Kelvin when so much was going on. It didn't help that it was winter and the days were short and dreary. I leaned heavily on my girlfriends for support. I often called Charley Webb, who had played Kelvin's daughter on *Emmerdale*. She has three children.

'How did you cope with the jump from one to two?' I asked her. 'It's crazy.'

'You can't think too hard about it,' she told me. 'You're in it. You've just got to live it. I know you can do it, Liz.'

Lucy Pargeter, another of Kelvin's former *Emmerdale* colleagues and a mother of twins, had been a great source of information when I first learned that Milo would be delivered by C-section. Like Charley, she cheered me on, calling and sending text messages that always seemed to arrive just when I needed them most. To have someone say 'I know what you're going through' meant the world to me.

The musical was finished by Christmas, thank goodness, but at the beginning of 2019, with our US visas still yet to come through, Kelvin was off again, this time to work on a charity performance of *The Full Monty*.

Right after that, Kelvin got a part in *We Go in at Dawn*, a Second World War film, playing the leader of a specialist team sent to break an Allied war planner out of a German POW camp. It meant two weeks in Suffolk (and growing a moustache), but it felt like a part he should take. The whole point of Kelvin leaving *Emmerdale* was that it would enable him to take his career in a new artistic direction, with more varied dramatic roles.

Meanwhile, I kept going, holding the fort, juggling motherhood and my own career, travelling to London to record voiceovers whenever I could. All the time I was just longing for those American visas to come through. It felt like only then could we really start making plans.

KELVIN

By this time, I'd been up for *Strictly Come Dancing* three times. On the first occasion, I met the production team and talked about the show, but I didn't get a call back. The second time, they'd asked me to show them what I could do on the dance floor. The result was another no. The third time, they hadn't even called me in but just asked Olly if I might be free around the time the show was running. And that was all. We didn't hear anything more.

A little later, Liz and I took the children on holiday to Lanzarote. You can imagine how disappointed I was to pick up a British newspaper one morning while we were there and see photographs of the new *Strictly* line-up – those celebs that the BBC had chosen ahead of me. Again.

I was gutted. I asked myself what it was that people *didn't* see in me. Was I deluded to imagine I'd ever had a chance? After three years of trying and three years of disappointment, I told myself that I'd done everything I could. *Strictly* was obviously never going to be for me.

I showed Liz the paper.

'That's it then,' I said.

'I don't know,' said Liz. 'I've got a funny feeling.'

She told me that she'd had a premonition that I would be on the show. Liz often gets weird feelings about the future. About unlikely things. They often come true.

'Don't, Liz, just don't,' I said this time.

'I just think you're going to do *Strictly*.'

'Well, obviously that's not going to happen,' I told her. 'They've announced this year's contestants. They're probably already in rehearsal.'

And if there was one thing I knew, it was that I was *never* going to put myself through another interview for *Strictly* again. There comes a point when you have to accept that you're pushing against a locked door. I didn't need that kind of disappointment.

*   *   *

A week or so later, we were back at home and Liz was changing Milo's nappy when the phone rang.

'Are you sitting down?' Olly my agent asked.

The tone of his voice suggested something big had happened. Something really big. For a moment, I fantasised that Marvel had been in touch. They wanted me to play 'Captain Britain'.

It wasn't Marvel.

'It's *Strictly*,' said Olly.

'What? Why are they calling?' I was immediately on the defensive.

'*Strictly* wants you.'

The line-up was already decided. I'd read all about it on our holiday.

'Somebody's been injured. They need a new celeb.'

I thought Olly was joking and it wasn't very funny. I told him as much. He insisted he wasn't kidding.

'They want you, Kelvin. Do you want to do it or not?'

I told Olly I'd have to think about it. He said I could have an hour to talk it over with Liz.

'Just an hour. The BBC are waiting on your answer.'

When the call ended I went over to Liz, where she was sitting with Milo in her arms. She'd been listening in.

'Are you going to do it?' she asked me.

'I don't know,' I told her honestly. I reminded her that three times I'd put myself up for *Strictly* and three times they'd turned me down. Now they wanted me, but only because someone else had had to drop out. I didn't like the way it made me feel. Like an also-ran.

Liz shook her head at me and smiled. 'Kelvin,' she said, 'shut up. Of course you have to do it. This is your moment!'

I continued to protest but Liz knew me better than I knew myself and she knew that despite the way I was talking, that call really was my dream come true.

'You're going on *Strictly*,' she said.

And that was that. Ten minutes later, I was back on the phone to Olly telling him to tell the BBC that I would be delighted to be on their show.

*　　*　　*

Liz and I found time to sneak in a weekend in Rome after that, knowing that once *Strictly* went on air, everything about our lives would change. *Strictly* was without a doubt the biggest show on the BBC.

Because I was joining late, I'd missed the matching process. All the celebs had their professional partners, so I knew from the day Olly called that I was going to be paired with Oti Mabuse, South African dance star and international Latin champion, for what was to be her fifth *Strictly* series.

When I met Oti for the first time, she looked me up and down and asked, 'How tall are you?' I didn't know until then how important it is for dancers to be matched in height. She was probably wishing she was still going to be dancing with Jamie Laing, the *Made in Chelsea* star who'd been forced to pull out with a foot injury. I felt bad for Jamie as I looked forward to the days ahead. I knew he must be feeling gutted about missing out.

There wasn't much time to dwell on how I'd got to be on the show once rehearsals started. I was thrown in at the deep end. Oti didn't waste any time putting me through my paces. As soon as I said yes to the show, Oti and her husband Marius sent me a video of the dance I was going to have to learn in a matter of days. It was the samba. And it was going to change everything.

CHAPTER NINETEEN
STRICTLY KELVIN

LIZ

The first two weeks of Kelvin's *Strictly* experience went by in a blur, with Kelvin and Oti practising late into the night. On the second Friday morning, he travelled down to London to rehearse the live show. I was going to be there for the real thing the following day.

Kelvin called as I was in the car on my way down on Saturday afternoon. I was surprised at how he sounded – like a boy again, completely unlike the confident man I knew. I realised I might have underestimated just how hard this was for him, how overwhelming it all was, with the intensive training and the crazy press. Even the costume fittings seemed gruelling.

'I wish you were here, Liz,' he said. 'I wish you were in the dressing room right now.'

He sounded like he needed a hug.

'I wish I could be there too,' I told him. 'You know I'm with you in spirit, though. And tonight I'll be right there in the audience. You'll be OK.'

Kelvin didn't seem convinced. I tried not to let my worry for him come through in my own voice.

That phone call was the last contact I'd be able to have with Kelvin until the end of the evening. For each show, Kelvin had two passes for the green room and two for the audience. That first day, I went along with Olly, our agent. As we guests arrived to wait in the bar, we had to hand over our phones, to make sure that no one was leaking exclusive gossip or pictures to the press.

When at last we were ushered to our seats, the atmosphere in the studio was electric. There's no audience quite so excited as a *Strictly* audience. As the familiar theme tune started, the cheers and whistles were deafening.

Then when Kelvin stepped out onto the dance floor, I couldn't believe my eyes. I'd got used to seeing him in his jeans or his Andy Sugden overalls, but here he was in a tight blue satin shirt, dark navy flares and Cuban heels. It was a total transformation.

'Is that really my husband?' I asked Olly.

Meanwhile, Oti looked amazing in a glittering orange fishnet dress and multi-coloured feathers.

They struck a pose on the podium and waited for the music to begin as the familiar voice of the *Strictly* compere intoned, 'Dancing the samba, Kelvin Fletcher and Oti Mabuse ...' Recognising as I did that little hint of uncertainty in Kelvin's smile, my heart was in my mouth.

'You can do it,' I told him silently.

From the very first note of the song – '*La Vida Es Un Carnaval*' by Celia Cruz – I could tell that Kelvin had it nailed. He looked confident and in control. More importantly, he looked as though he was really enjoying himself. He was putting into practice the advice he'd always given me: that if you were going to do anything, you had to do it with conviction. You had to go in 100 per cent.

'If you don't own it, it shows,' Kelvin would say. 'You've got to embrace what you're doing and give it your all.'

He was walking the walk. Or rather dancing it! I couldn't believe how amazing he looked out on the floor – and he only got better and better.

The audience went wild! Jamie Laing was sitting nearby. It must have been a bittersweet moment for him, seeing Kelvin take to the floor for what should have been his first dance. But he applauded Kelvin's performance harder than anyone.

When Kelvin and Oti took their bow to a standing ovation, Olly turned to me and said, 'Well, that couldn't have gone any better.'

It wasn't just the audience that stood up for Kelvin that day. Three of the judges got to their feet to applaud him.

Judge Motsi Mabuse, Oti's sister, said: 'Kelvin. Are we not happy you stepped in? You have good assets,' she added, sending the audience into roars of delighted laughter. 'I'm talking about that ooziness in your body ...'

Meanwhile Shirley Ballas fanned her blushes and said, 'I'm the oldest here so, darling, I need my fan and my beads, but never mind ... You are outstanding.'

Bruno Tonioli agreed: 'This hunk is on fire! I think you started a chain reaction of hot flushes ... This is the best samba I have seen on a debut show.'

Craig Revel Horwood was harder to please, of course, beginning his comments with criticism. 'Your hands bothered me ... the bounce absolutely needs work ... but God works in mysterious ways. Amazing.'

Backstage, Claudia teased Kelvin: 'You should only ever have two weeks' notice for everything you do.'

Then the scores came in. All of the judges gave Kelvin and Oti an 8 – even Craig – making a total of 32, putting them at the top of the leader board.

The viewers were delighted too. Kelvin and Oti's samba was the dance that got people talking. Twitter blew up. Everyone was going crazy. Kelvin and Oti had broken the internet! It was a matter of hours before the first GIFs started appearing. People were asking, 'Who is that?' On Twitter, one fan wrote, 'Well Kelvin IS a unit of heat.' Someone else compared him to Patrick Swayze.

Kelvin's score wasn't just the best that night; it was the best opening score on *Strictly* ever. It even inspired a section on Channel 4's *Gogglebox* later in the weekend.

It was a brilliant start and the next day we were still on a high. But it was just the first dance. The following Monday morning, Kelvin was starting from scratch again, with a new dance, a new costume and a new challenge for the newly christened #teamfloti.

KELVIN

I'd always known that the moment I started dancing would be the moment of 'make or break'. Most *Strictly* viewers, if they had heard of me at all, had a fixed idea of who I was. I was Andy Sugden, fictional farmer, wearer of overalls and keeper of cows. What they didn't know was that behind Andy Sugden was a guy who had been brought up on Motown and soul. I had that music in my heart. I was invisible until I danced.

That first live show was mind-blowing – I couldn't have imagined a better start – but almost as soon as the night was over, it was as though it had never happened. The following day, back in Oldham, I was just plain old Dad again. Children are great at cutting through the BS. It's hard to get above yourself when your baby son needs his nappy changing. Likewise, a high score from a *Strictly* judge means nothing to a toddler who wants her breakfast *now*.

And on Monday, it was straight back to the grindstone. Oti and I started rehearsing the next dance at nine o'clock sharp. We did most of our rehearsals at a studio in Manchester, so that at least I could stay at home during the week. It was great to be able to spend my evenings with Liz and the children. But rehearsals were no holiday. They were physically gruelling and really hard work.

At the beginning of each week, Oti would demonstrate the dance that I was going to master over the next five days. Oti choreographed all our dances apart from the Charleston and the salsa. She always made it look effortless but within five minutes of trying the steps myself, I would realise just how difficult the dance really was and my heart would sink.

Mondays, Tuesdays and Wednesdays were always an uphill struggle as I tried to commit the dances to memory and master steps I'd never done before. Thursdays were usually the game-changer. That's when I'd feel like I was finally beginning to get it and adding some flair. Then on Thursday evenings we would travel down to London for rehearsals for the live show. On Friday we'd rehearse the running of Saturday night, record 'snippets' to be played between live dances and get notes from the producers. By ten in the evening, Oti and I would inevitably still be in the rehearsal room. Oti always wanted me to give it my best on a Friday so that she could gauge whether we needed more. I'd send Liz the rehearsal videos, then call her late at night, full of worry that this was going to be the week when I made a complete fool of myself.

'You'll be all right,' Liz told me every week.

It's to Liz's eternal credit that she let me have my moment of panic like that every Friday night like clock-work, while she was holding the fort back home and doing her own work as a busy voiceover artist. To say that she had her hands full is an understatement. As stressed as I might have felt the night before the live show, I was living the *Strictly* dream. Liz was dealing with the nappies, the night feeds and the terrible twos, and she was doing it all with her usual unflappable grace.

In *Strictly*, you're in a bubble. You're in training all day every day, until nine, ten, eleven at night. It consumes you. I was totally immersed in dancing. Even while I was eating dinner, I'd be watching dance videos at the same time. I wanted to know everything about it. I loved seeing the professionals practise their group routines for the live show. I was in awe of them. The way they danced when they were unleashed from us amateurs was something else. That's when you saw their real skill.

I was glad I had Liz and the children to come back to every Sunday. The comedown after the live show was always tough but the children helped me to keep my feet on the ground. I did take Marnie to the studio from time to time and she loved it. Once she was on the dance floor, it was hard to get her off. The other dancers – celebs and professionals – were very kind to her, especially Anton Du Beke, whose daughter is around the same age. Marnie couldn't get enough of him. Every time Anton had a spare moment during rehearsals, Marnie sought him out.

Even Dad started getting into it. His face when I told him that the producers had decided the 2019 series would include a show in which the celebs danced with their parents was a picture.

'You'll be standing next to Tess,' I told him. 'She'll introduce you.'

It was just a joke and I didn't really expect him to fall for it, but he did. Dad got his hair cut and bought some talc, to help him glide across the floor in his dance shoes. I had to come clean when he told me he was booking a day off work to be at the studio.

I'm pleased to say that Dad got his own back when he made a guest appearance on *Strictly*'s sister show *It Takes Two*, dancing into the studio with some impressive Northern soul moves that had the crowd in an uproar. I had no idea he was going to be on the show. When he sat down next to Zoe Ball, he told me, 'That is how a proper prank is done.'

It was soon hard to imagine life before *Strictly*. It felt as though dancing was my life. At the same time, every week I prepared for it to be my last. Even after Oti and I scored a perfect 40 for our quickstep to 'The Lady Is a Tramp', I didn't take it for granted that we'd ever get that close to perfection again.

In many ways, it didn't feel like a competition. Everyone just wanted to dance as well as they could. And no matter how well the judges thought Oti and I had danced, the ultimate decision lay with the viewers. Who would they vote for? What would they be judging us on? Not purely on our dances, was my guess, but on who they thought we were away from the camera. Or maybe even who they thought we were when we were playing the fictional characters they loved.

And yet, I made it to the final show, alongside Emma Barton and Karim Zeroual. It was to be the first three-couple final since 2016, as Will Bayley, the Paralympian table tennis champion, had had to withdraw from the competition early, leaving the show a contestant short. He'd landed awkwardly when leaping from a raised platform during rehearsals, ending up with a leg injury. We were all gutted for him. During the show's run, I'd often thought about how hard it must have been for Jamie Laing to have to miss out on his shot at *Strictly*. For Will to have got so far, only to have to drop out right in the middle of the series, must have been really painful.

In preparation for the big day, the finalists each had to record a segment in which we addressed the audience at home directly, asking for their vote. I didn't know how to approach it. It cost money to vote. I didn't feel like I could ask so shamelessly. Who did I think I was? Instead, as I sat in front of the camera, I decided I wanted to tell the audience how thankful I was for everything I'd experienced.

'I'm having an amazing time here …'

I knew I'd already lost one vote. Days before the final, Marnie told me matter-of-factly that she would be voting for Emma Barton and Anton Du Beke because they were 'a real-life prince and princess'.

CHAPTER TWENTY

BRINGING HOME THE GLITTERBALL

LIZ

Kelvin's time on *Strictly* made special memories for the whole family. Though Milo was a bit young to know what was going on, Marnie adored watching her dad on the show – she'd get really close to the screen and point at him every time he was on.

Then suddenly, twelve weeks had passed and Kelvin was through to the *Strictly* final. I was so pleased for him. What the viewers at home see on a Saturday night gives no real sense of the hard work and heartbreak involved in every dance. Even if Kelvin didn't win, I knew he should be very proud of how far he'd come

The final show was a big family affair. In Kelvin's corner there was me, his mum and dad, his two brothers and our good friends George and Joanna Burgess. As a finalist's family, we were given seats in the front row and a constant stream of people came up to us to tell us how much they were looking forward to seeing Kelvin dance that night. For the *Strictly* final, the production goes up a level and we couldn't wait to see what the team had pulled out of the bag.

I knew from talking to Kelvin earlier that day that he was nervous but also in the zone.

'I feel like tonight could go any way at all,' he told me. 'Any way at all. But I am ready.'

Though Ladbrokes, the betting company, had him as the favourite by this time, Kelvin wasn't convinced. He didn't believe he was that popular with the audience, though I'd heard on the grapevine that at the Blackpool show midway through every *Strictly* series they have a clap count, measuring the decibels of the applause, and that year's audience reaction to Oti and Kelvin was the loudest ever.

Kelvin was dancing against Karim and Amy Dowden and Emma and Anton. As it was Anton's seventeenth year on the show and he had never yet won the glitterball, I think a lot of people were expecting him to win.

'Just concentrate on the moves,' I told Kelvin, just as I told him every week.

And at last, it was time for the show to begin.

KELVIN

Standing backstage before the show that night, I closed my eyes and imagined the dances ahead of us. Oti and I had put in the hours. We couldn't have worked any harder. With all the preparation we'd done, all we could do now was hope that the dance gods were with us.

We'd prepared three dances for the final. A special show dance to the Isley Brothers' 'Shout', on a set like a giant jukebox, our week-four rumba to Bill Withers' 'Ain't No Sunshine' and, of course, the samba from week one that had been such a hit with everyone – even Craig Revel Horwood. Would it be such a crowd-pleaser second time around?

Hearing the cheers as our samba was announced, my heart was immediately lifted. It was the dance everyone wanted to see.

'Ready?' Oti asked me.

I nodded. I was as ready as I ever would be. We struck our pose on the podium and waited for '*La Vida Es Un Carnaval*' to begin.

It was magical to be on that dance floor. The atmosphere in the studio was off-the-scale fantastic. When you're dancing, you can't really see the audience; you certainly can't focus for long enough to pick out individual faces, but I could feel Liz and Mum and Dad and my brothers in the audience, willing us to ace every move. I was dancing for all of them.

We scored 39 for the rumba, another 39 for the samba and a perfect 40 for the show dance. They were amazing scores but not quite amazing enough. Karim and Amy pipped us to the top of the leader board by a single point, thanks to their two perfect 40s. But though the judges had chosen their best dancer of the night, we all knew that the title of *Strictly* champion was still in the gift of the millions of people out there watching at home. Who would be their winner?

When at last the results were in, we lined up to hear the public's decision. Would they agree with the judges, that Karim and Amy's technically perfect performance was the best of the show? Or would the huge fan following Anton had gathered over the years see him and Emma take home the glitterball instead? I felt sure of only one thing as I stood there: it wouldn't, it couldn't, be me and Oti …

LIZ

Sitting in the audience, watching Kelvin stand there waiting for the results, my heart went out to him. I knew he had given those three final dances everything.

Tess and Claudia began to talk, dragging out the big moment for as long as they could. Me and Kelvin's Mum and Dad held hands as Tess said, 'The votes have been counted and independently verified, and I can now reveal the *Strictly Come Dancing* champions are …'

In the pause that followed the audience clapped and time seemed to stretch out forever until Tess yelled out, 'Kelvin and Oti!'

Kelvin's face was a picture. I could tell he really was surprised. As he disappeared into a knot of hugs on the stage, I hugged his mum and dad. Warren was in tears. 'I'm in bits,' he said when he was able.

On the dance floor, Tess handed Kelvin the microphone.

'I am absolutely speechless. I did not expect that,' he began. 'It has been such a privilege to be here. I think this show represents everything that's amazing with this country. I think the people personify what is great and it's just been an absolute privilege …'

He was so emotional he couldn't finish the sentence but burst into tears instead.

When he was able to continue, he thanked Oti and the other finalists and the whole production team. And 'those at home – everyone who has supported me … I'm going to go and cuddle my wife, cuddle my mum and dad, and my brothers are here. I think I need five minutes just to take stock really. It still feels just so surreal, and when you don't expect something and it happens it's just really quite hard to take. I wish I had more words in my vocabulary to explain how I feel.'

It wasn't only Warren who was in bits by this point. I was feeling pretty emotional too. We were all so proud of our Kelvin. From last-minute stand-in to champion, he had gone all the way.

KELVIN

The moment when I was announced as the winner was totally surreal. I don't think I really understood what was happening at the time. Looking back at the video of the announcement, I can see my jaw drop in complete and genuine disbelief.

I wasn't supposed to have got this far. I wasn't even supposed to have been on the show. I'd started out as the underdog and somehow I'd become the nation's favourite. My journey to this moment was like the script of a Hollywood movie.

Oti was over the moon. It was the first time she'd won *Strictly*. The following year she would win again with Bill Bailey, making her one of only two professional dancers to have lifted the *Strictly* glitterball twice.

When you learn that people have voted for you, it's a really lovely feeling. Every time you're saved by the public vote, it's humbling too. I was stopped in the street by people who told me I had their vote. I can't think of any other job that gives you such a high. I was grateful beyond words.

It was one of the best nights of my life. I couldn't have been more glad that Liz had seen through my hurt pride after I told my agent Olly I'd have to 'think about it' when he called to tell me that *Strictly* wanted me after rejecting me three times. Thank goodness she knew me well enough to tell me to stop pouting and seize the opportunity with both hands. Starting with Liz and Oti, I had so many people to thank. Not least, Jamie Laing, whose place I had taken. On social media, he posted, 'Congratulations to @kelvin_fletcher and @OtiMabuse you guys rocked it!! Thank god for my broken foot.'

I was more grateful to him than he'd ever know. As were several of my mates …

Although Ladbroke's had me down as the favourite in the final, at the outset my win was so unlikely that a few of my friends had done very well by putting bets on me before the series began. My brother got odds of 16–1, which translated into a nice pay-out. One of my friends benefited from my win to the tune of £17,000!

I still have the glitterball, of course. It's much smaller than you imagine when you're sitting at home seeing it on the television screen, but it's fabulous all the same. Now it sits in our lounge on the farm, reminding me of life before we had sheep.

CHAPTER TWENTY-ONE

LIFE ON HOLD

KELVIN

The days after my *Strictly* win were a whirlwind of interviews and TV appearances. When that all died down, Liz, the children and I took a well-earned family holiday in the sun. Then as soon as we got back home, I started rehearsing for the next job, which was the *Strictly* arena tour.

Liz had her own work to do. While I'd been doing *Strictly*, she'd been cast in the new series of *Cold Feet*. Her scenes were with Robert Bathurst, who'd played Hermione Norris's husband in earlier series of the show. Liz, of course, got her first TV acting break opposite Hermione Norris. The acting world can sometimes feel pretty small.

The *Strictly* arena tour, which kicked off in January 2020, was a great bonus. I couldn't wait to get dancing again and it was a chance to really get to know Craig Revel

Horwood, who directs the tour, and the others. During the TV show, we'd only come together on Fridays and Saturdays and didn't have much time to mingle and chat. On tour we were all together away from home. We'd be in Belfast having a drink and at last we could drop our television personas and be ourselves. Needless to say, Craig is much nicer than the tough judge you see on screen.

While all this was going on, our US visa applications were still chugging through the system. We still assumed that 2020 would be the year we moved to the States. But then we received a letter from the US Embassy saying that for the time being all visa applications had been put on hold. By this time it was February 2020. We'd heard about the new virus that was causing trouble in China but never thought it might affect us in the North of England.

With our visas on hold and no idea what was coming next for the whole world, Liz and I decided we needed to make a new plan. We had put our house on the market, expecting to be moving to the States. We decided that we would go through with the sale and rent in the UK for a while. But where? Perhaps it was time to dust off a dream.

LIZ

Kelvin goes on Rightmove like I go on Instagram. He's addicted. Since the site has existed, he's had a big 'favourites' list of dream homes that we could never hope to afford: like the Scottish castle near Inverness and the London house on the market for £43 million. I suppose it was his property equivalent of my NET-A-PORTER wish list full of designer shoes.

But one day, Kelvin told me about a place that wasn't quite so far out of reach. While travelling down to London for *Strictly*'s live show, he'd looked out of the train windows near Macclesfield and liked the look of the beautiful countryside passing by. When the train stopped at the station, Kelvin decided to have a look on Rightmove to see what kind of property was in the area. And that is when he found himself looking at a 120-acre farm.

'A farm?' I exclaimed when he told me.

I had to agree that Macclesfield seemed like a nice place to live; the countryside was beautiful, Macclesfield itself was a decent-sized town where you could get everything you needed, and it had convenient train links to London and Manchester. But the thing was, we weren't even looking! We were supposed to be getting ready to move to the States. And now Kelvin was talking about a farm. I knew he'd spent twenty years on *Emmerdale* but that didn't exactly qualify as farming experience, did it? As for me, I was definitely a dyed-in-the-wool townie.

Still, Kelvin told me he 'had that feeling'. And who was I to tell Kelvin that 'having a feeling' about something wasn't to be taken seriously? I'd had a feeling that he would be on *Strictly*, after all.

The last time Kelvin had such a strong feeling about a house was when he was just a kid. When we were growing up in Royton, an estate of five big, detached houses went up nearby. The houses were surrounded by a big stone wall that added to their exclusivity. Everyone tried to get a glimpse over that wall. The houses were the most expensive the town had ever seen. The people who lived in them must be beyond rich. Kelvin decided that one day he was going to be among them.

It just so happened that one of those houses came up for sale in 2008. It was the first time one had come on the market since they were first built. By then Kelvin had been in *Emmerdale* for years, putting money aside, and finally he could afford one. That house became the first home we shared together. It was the house we were getting ready to sell for our move to the States. We still hoped that our US visa applications would eventually reach the top of the pile but Macclesfield was a very long way from LA.

All the same, I let Kelvin tell me more about the farm and agreed that we could at least go and have a look. But by the time *Strictly* was finished and we had a little breathing space before the start of the Live Tour, we'd missed our chance. Kelvin's dream farm was already under offer.

I thought that would be the end of it, but Kelvin wasn't to be put off. My mum and dad had recently moved to the countryside near Chester and they thought it would be great if we moved closer to them. On the way back from visiting them one weekend, we drove past Kelvin's Rightmove dream farm and took a closer look at the area.

A little while later, when Kelvin had a day off and the children were at school and nursery, he persuaded me to go for another drive near Macclesfield. Kelvin had Rightmove open on his phone and as I drove, he looked for properties for sale nearby. He soon found another farm. We drove right up to the gates to have a nosey. Liking what he saw, Kelvin decided we should knock, and, in his inimitable way, he charmed the owner into letting us have a look around there and then, though we didn't have an appointment.

It was an amazing place, really beautiful, but not right for us. It did, however, convince Kelvin that his idea that we should buy a farm was not just a fantasy.

'We could do it, Liz. Can't you imagine us living in the countryside? The kids growing up here?'

I could tell that he was getting serious. When we got home that day, Kelvin got a phone call from the agent who was selling the farm we'd visited, asking if we were interested in seeing other properties. Kelvin told him about the farm he'd actually fallen in love with, the one which was under offer. The agent promised to call if that situation changed. I was secretly relieved to hear the farm was off the table.

I wasn't yet convinced by Kelvin's new farming dream. Though for some time we had been subconsciously planning a move – we'd kept our house decorated in a really neutral way, all grey and white, so that it was ready to go on the market – we'd never talked about making a move to the country. The US plan seemed much more sensible. We were both actors. We liked California. We loved the weather, the people, the vibe. I'd assumed that while we waited for our visas, we would stay put, near Manchester, near our friends, enjoying the sort of city lifestyle we'd built up over so many years.

I loved where we lived in Oldham. I loved being close to all my friends. I loved the idea that our children would have the same kind of childhood as Kelvin and I'd had. They'd go to a local school and have lots of mates nearby. They'd be able to play out on the street, just as we'd done.

All the local adults would keep an eye out for them, just as we looked out for our friends' kids. I wanted that sense of community. Now Kelvin was proposing to move somewhere really isolated. What's more, what we knew about farming we could write on the back of a stamp. Half a stamp. Hard as I tried, I really couldn't see the benefits.

But if there is one thing I know about Kelvin, it's that when he sets his heart on something, he makes it happen. Over the next few weeks, we had many conversations about the pros and cons of leaving the neighbourhood where we were so established. Gradually, I came to realise that in idealising my own childhood and wishing that my children could have exactly the same, I was overlooking the many advantages the country life might have to offer them. And us. The truth was that we had no idea when or whether we'd get those US visas. Perhaps it was time to think again.

Some of Kelvin's arguments for moving to a farm were more successful than others. I liked the idea of the children being able to safely ride around a farmyard on their bikes or build treehouses. I also liked the thought that they might get to grow up alongside animals. I was less swayed by 'If we had a field, we could walk around naked whenever we liked!' All the same, it wasn't long before Kelvin had me persuaded and I was the one saying, 'Let's just do it!' For most people, making such a big decision would have required more research, more exploration, but we've always had the feeling that the best decisions come from the heart.

Then fate stepped in again. The estate agent selling Kelvin's dream farm called to let us know that it was about to come back on the market. It had been under contract to an entrepreneur with an events company, who'd seen its potential as a venue for festivals and concerts. But that was before he had heard the news of the virus, which was, by now, making its presence felt all over the world. Kelvin had already been told that a cruise on which he was to make a guest appearance had been cancelled. We got the impression that people in the events industry had some kind of advance knowledge about what would happen everywhere in March.

'Do you want to put an offer in?' the agent asked.

Of course we needed to see it first. Kelvin said we'd be there as soon as we could. He didn't even change out of his Fitflops and shorts. The owner of the farm and the very posh agent must have wondered what they were dealing with, as Kelvin traipsed through the mud of the farmyard on a chilly spring morning with his toes out.

That day, I was booked to do a voiceover, so I only had time for the briefest tour before I had to catch a train down to London. I left Kelvin behind with the agent and owner. He stayed on the farm for another three hours that day, making sure he knew exactly what was for sale and what its potential might be.

By the time we met up at home that evening, Kelvin was ready to make an offer.

'I can see us living there, Liz,' he said. 'I've got that feeling again.'

Kelvin's enthusiasm wasn't just based on a whim. He explained to me that the farm would be more than a wonderful family home. It could be a source of earnings too. He explained how a large part of the farmland was already rented out to other local farmers and was generating a regular income that would offset some of the costs. As actors, we both knew how important a regular income was.

Suddenly, the idea of owning a farm became more and more attractive. We felt like it would give us some control when our acting careers were going through a quiet patch. Remembering how it had been when Kelvin first left *Emmerdale* and he couldn't seem to catch a break, the thought of diversifying so that we didn't have to rely entirely on acting to pay the bills was comforting.

While Kelvin and I weighed up the future, the world was coming to terms with the fact that this Covid-19 virus was not going away quietly, despite what certain politicians would try to have us believe. On 23 March 2020 Boris Johnson announced that the UK would be going into lockdown. That same day, we put in an offer on the farm. When we put the phone down, we squealed with joy. It was a strange moment. We had just taken a concrete step towards creating our future when all of a sudden the future of the entire world looked very uncertain indeed.

That first lockdown was a surreal time. Everything ground to a halt. I know it was frustrating for Kelvin that all the opportunities that came about as a result of his winning *Strictly* were on hold. Talk about bad luck. All the same we felt lucky that we had a garden and the spring weather was so beautiful and warm. We went for our daily walks and fired up the barbecue. There was a real sense of stillness, which was good for us in many ways. It gave us time to reflect on our decision to buy the farm and prepare for the transition to a very different way of life – and, confined to our small area of Oldham by the lockdown rules, I found myself longing for the wide-open expanses of the countryside.

When we put in our offer on the farm, Kelvin had assured the agent that we would be able to move quickly in order to get our offer accepted. Of course, lockdown put a stop to that. The process was extended indefinitely. That was not such a bad thing. Both Kelvin and I had bought houses before, but we soon realised that buying a farm was absolutely nothing like buying a two-bed terrace in a city centre. Lockdown gave us the time to research the intricacies of buying farmland – and they were very intricate indeed.

As is his way, Kelvin put 100 per cent into understanding everything there was to know about 'entitlements'. Sometimes he would be up until two in the morning researching some minor point of law. When the estate agent produced the file on the farm, Kelvin read every single word, checking anything he didn't understand with his brother Dean, who is a lawyer.

The minute estate agents were allowed to show prospective buyers around again, we put our house on the market and it sold right away. We still weren't ready to complete on the farm, but we didn't want to lose our buyers so we arranged to move into a rented house in Alderley Edge. Though the house we rented was much smaller than the house we'd sold, we loved it. It was convenient for Marnie's school and we enjoyed the cosiness for a while.

During those periods when lockdown was eased and Kelvin was able to travel, he took on a couple of jobs. He did a drama in Liverpool (*Moving On* by Jimmy McGovern), then – lucky thing – went to the Caribbean to film an episode of *Death in Paradise*. Meanwhile, I kept up my voiceover work, recording regular ads from the cupboard in Marnie's bedroom! We all mastered Zoom. Together with my friend Helen Rigby, I made plans to record a podcast, which would become Soul Story Club.

All the time, our excitement about the farm kept building. We'd received another email from the US embassy, telling us that our visa application was likely to be on hold for another eighteen months as the States got to grips with Covid. It seemed like a sign. Our move to the countryside was meant to be. Having the option of moving to the States taken out of our hands made it so much easier to put that dream out of our minds and focus wholeheartedly on the farm.

*　　*　　*

We spent Christmas 2020 in our little rented house in Alderley Edge, making the most of the quiet time together. Like most people up and down the UK, we'd seen our Christmas plans upended by the new Covid variant that led us into a second lockdown in early 2021.

By January 2021, we were sure that it must be a matter of weeks before we were able to move to the farm. Among the many Zoom calls we made during the third lockdown, Kelvin had a chat with a producer at the BBC. Kelvin's *Strictly* win and the fans he'd picked up along the way made the BBC keen to use him in something else. During the brainstorming chat, they came up with documentary ideas. Perhaps Kelvin could do a sort of engineering show, celebrating the North. Then the producer asked why Kelvin was dialling in with a backdrop of boxes.

'We're moving,' Kelvin said. 'We've just bought a farm.'

CHAPTER TWENTY-TWO
BETTING THE FARM

KELVIN

The producer on the other end of the call was incredulous. 'You mean to say you literally just saw a farm while you were on a train to London and decided to buy it?'

'Basically, yes,' I told her.

'That's hilarious,' she said. 'And also absolutely crazy.'

But I could tell that an idea was forming in the producer's mind. This was a lightbulb moment. I'd played a farmer on *Emmerdale* and now I was going to do it for real? Had I really bet everything we had on a farm without a clue as to how it would turn out? It had all the elements of a sitcom. Or a documentary about modern farming. Since the first lockdown, everyone was more interested in the simple country life.

'Kelvin,' said the producer, 'I think we might be onto something.'

*   *   *

Very quickly, a plan started to come together. The only problem was, Liz and I still didn't actually own the farm I'd so lovingly described. I felt a sense of rising panic as I faced the possibility that the BBC might actually commission this show, only for me to have to tell them that it was all off because we didn't get the farm after all.

The buying process had been dragging on for almost a year, as the estate agent and the owner dealt with issues regarding the tenant farmers and agreements that had already been made to sell off pockets of land in separate deals. It was complicated, to say the least. I didn't know what I would do if it all fell through at this stage. Not only would it be the end of our dream, but I was now in danger of leaving an important producer feeling that I'd messed her around.

The relief when the agent finally called to tell me that the owners of the farm were ready to exchange contracts was immense. I immediately phoned the producer. 'We've got it,' I said. 'We've got the farm.'

'Yes, I know that,' she said, and I realised that she had been under the impression that it was already a done deal! Talk about a close shave.

* * *

As always seemed to happen every time we were about to move house or have a baby, Olly called a few days after we exchanged contracts to tell me that I'd been offered a job. This time it was a part in a psychological thriller called *The Teacher* opposite Sheridan Smith. The four-part drama was set in the UK but for various reasons it was going to be shot in Budapest in Hungary. Filming was going to start right when we were due to pick up the keys to the farm.

The timing wasn't ideal, but I had to say yes. The chance to work with the Channel 5 drama team was something I didn't want to miss out on. Liz agreed that I should do the job and assured me that she would be fine on her own with the kids at the farm while I was away. At least, it turned out, I could be there for the move itself.

The day we got the keys, it was snowing, which might have made the farmyard look romantic, but it really was the last thing we needed – not least because we very quickly found out that the ground-source heat pump, which heated the main house, had packed up. What perfect timing. It was freezing and I was off to Budapest in a matter of hours. Liz was her usual calm self as we unloaded her designer shoes and my *Strictly* glitterball, but she must have been wondering what on earth she'd signed up for: the trappings of our Oldham life looked distinctly out of place in this old house that was like an icebox. An icebox full of spiders.

But there was no turning back now. The BBC show was going ahead. Though it was going to be based on the farm, the producers knew that we had no plans to be proper farmers ourselves yet, so the rough idea was that I would act as a presenter, introducing the viewers to the countryside and the various interesting people in the locality. We met one of those people on our very first day.

While we were unpacking boxes, I'd noticed someone drive through our farmyard on a quad bike. Shortly afterwards, she came back again. This time she stopped to introduce herself.

'I'm Gilly,' she said. Her smile was warm and welcoming.

Gilly explained that she was the farmer who was renting some of our new fields for her flock of several hundred sheep. She'd been out on her quad bike to check on them.

'You're the new owner then?' she asked.

Beyond knowing that Liz and I had just bought the farm, Gilly had no idea who I was. She didn't watch *Emmerdale*. She was too busy farming in real life!

Gilly and I clicked straight away and had quite a long chat. At the end of it, she told me that if Liz and I ever needed any help or advice, we should give her a call. She was just a couple of fields away. It was comforting to me to know that someone so kind and friendly would be close by for Liz while I had to be away in Hungary.

By the end of moving day the heat pump was fixed, and with all the removal boxes unpacked (or at least unloaded), I decided that Liz and I needed to take a moment to celebrate our big move. I persuaded her to take half an hour off, leave the kids with their grandparents and drive around our land in my car. She'd been so busy with the move, she hadn't really had a chance to look around and appreciate what we had.

'Is that our field as well?' she asked, as we drove up to the top of the hill. 'And this one?'

I don't think until then that Liz really understood just how much farmland we'd actually bought!

I parked up at the top of our highest field and we sat in silent appreciation of the view. A spectacular sunset made it even more beautiful. I felt proud as we looked out over the fields below – ours for almost as far as we could see.

What a couple of years we'd had. We'd been pulled in so many directions and made so many plans that had been derailed by the pandemic. This was our life now. The countryside. The small local village with its church and its pub. The sound of the animals in the fields.

'We did it,' I said, pulling Liz in for a hug.

*　　*　　*

After those precious moments of reflection, it was time to go back down to the house and carry on with the unpacking. I turned the key in the car's ignition and made to drive off. Except that we couldn't. The car's wheels spun in the mud.

To begin with I just kept revving the engine, but after twenty minutes I'd succeeded only in getting us more firmly entrenched. I tried everything. I even got an ironing board out of the back of the car (still there from the move) and tried putting that under the wheels. Result: one ruined ironing board. One car still absolutely stuck.

'Kelvin,' said Liz, 'we've got to get some help.'

There was no point trying to get Dad or my father-in-law to pull us out. Their cars weren't four-by-fours. They'd only end up getting stuck too. Would a local garage send a tow truck out to a field in the middle of nowhere? I doubted it.

In desperation, I rang Simon, the former owner of our farm, and asked him for Gilly's number. I was too embarrassed to say why. But when Gilly picked up, of course I had to tell her what was going on.

'It's Kelvin,' I said. 'We met this afternoon.'

'Oh hiya, Kelvin. You're settling in OK?'

'You could say that,' I told her. We were settled in good and proper in the mud. I told her our predicament. To Gilly's credit, she didn't laugh.

'Which field are you in?' she asked.

To my further shame, I had no idea. That's right, I didn't even know which of our own fields we were sitting in.

'Describe it to me,' Gilly said. 'What can you see from where you are? Can you see any sheep?'

Gilly, who knew the land better than the back of her hand, soon had it worked out.

'Don't worry,' she said. 'We're used to getting people unstuck.'

Gilly sent her husband, Jack, out to rescue us. He arrived on his big tractor, with his apprentice sitting by his side in the cabin. It was the first time Liz and I had met Jack. I told him I wished this first meeting was taking place in the pub. Jack just nodded. Had I upset him? Had we dragged him away from his tea?

'Right,' he said, getting straight down to business. He and his apprentice assessed what needed to be done and hooked up the car to Jack's tractor. A couple of minutes later, our car was back on the road. Liz and I thanked Jack profusely, but his only response was a brusque, 'See you later on then.' There was no chit-chat. As I drove our car back down to the farmyard, I felt hot with embarrassment at the thought of how Jack or his apprentice might describe what had happened down the village pub. Two actors, playing at being farmers, getting their Jeep stuck in their own field. We'd be a laughing stock.

Since then, I've got to know Jack much better and realised that, like my dad, he doesn't waste his time on small talk or gossip, saving his breath for the conversations that interest him instead. Meanwhile, I've got much better at knowing where not to take my car when the weather's been really wet.

CHAPTER TWENTY-THREE

A PROPER (PEA)COCK-UP

LIZ

It seemed like planning to move house cast some kind of spell on Kelvin's career. Whenever we were ready to up sticks, he would get a job that meant I would be left to settle into our new place alone. But never before had I been faced with such a challenge as I was at the farm.

The day after we moved in, Kelvin was off to Hungary and I was left behind with the spiders. And there were a lot of spiders. Every single spider in the Peak District seemed to have moved into our new house for the winter.

Every day was busy. The children had to be taken to school and nursery. Making sure they were settled was my number-one priority. I quickly worked out new routes to get to the places I had to go each day. Once the children were where they needed to be, I'd come home and unpack boxes. But it wasn't just the house that needed my attention.

Those fields that had looked so picturesque as we watched our first sunset at the farm weren't going to take care of themselves. Though we didn't have any livestock of our own as yet, many of our fields were rented out and did contain sheep or horses. It was our responsibility to make sure that the fences and walls around those fields were maintained. That meant I had to walk the fields every day, checking for damage. Later, we'd get a quad bike to make the job easier, but in those early days, I had to do it all on foot and often with the children in tow, holding Marnie by the hand and carrying Milo. It was hard work but at least it meant I didn't miss the gym back in Oldham.

Every evening, Kelvin would call from Budapest. He told me that working with Sheridan Smith was an amazing experience. The only sad part was that, due to Hungary's pandemic restrictions, when they wrapped for the day the cast and crew weren't allowed to socialise with each other, but instead had to go back to their hotel rooms and spend the evenings alone. How sad we both were: Kelvin on his own in Budapest and me on my own in the farmhouse, after the children had gone to bed. Ginger the dog wasn't much company. She was usually flaked out after a busy day investigating farmyard smells (and bringing them inside on her fur).

It was tough, but I had to believe it would be worth it. Looking out of the window at the farmyard on a cold winter's day, I reminded myself that before I knew it spring would come, and when it did our children would be able to play outside, riding bikes and building tree-houses in our woods. We'd never regret giving them the chance to have those adventures. In the meantime, so long as everyone was healthy and happy, we were doing OK.

*   *   *

Soon Kelvin was back from Budapest and the plans for our farm show picked up pace.

The television crew arrived in May – the director, the producer and two camera operators. After those early weeks when I was on my own with the kids, it was good to have a few more people around the place. The crew quickly became part of the family.

As the start of filming approached, we knew to be mindful of the fact that we'd only just moved to this part of the world and we didn't want to upset people in our new village. We reached out to them and let them know what was happening. Most of the people we spoke to seemed OK with the idea of the show. They understood that the film crew would be very small. Kelvin and I weren't the Kardashians. That said, it was going to be a new way of living for us, going about our business with a cameraman in tow.

It was the first time Kelvin and I had worked together since *Bugsy Malone*! However, this was no school play. Neither was it like any play, or film, or episode of a soap we'd ever been involved with. As actors, we were used to things happening according to a script, but there was no script now. It would have been foolish to try to write one. There's only a certain degree of planning you can apply to life on a farm. The cameras might have to roll all day long in the hope of getting just a few moments of useable footage.

We had a long to-do list of things we wanted to get done around our new home, and having a TV crew filming us every day meant that we were soon ticking things off that list, in order to give them something to film! We shot hours and hours of footage but an awful lot of it would not make the cut – like our very first venture into owning livestock.

Viewers of *Kelvin's Big Farming Adventure* might think that the first livestock we came to own were our Cotswold sheep. That's not strictly true.

It started while Kelvin was still in Budapest filming *The Teacher* and I was alone on the farm. One of the things I wanted to sort out was having someone in to clean all the farmhouse windows, which hadn't been done in a while. When he'd finished the job, the window cleaner stopped for a chat and said something I never thought I'd hear.

'What you need is a couple of peacocks.'

'Peacocks?' I asked.

'Yeah. Can't you just imagine it? You've got that lovely lawn there out the front. Your visitors will be driving up and the first thing they'll see is a magnificent peacock strutting his stuff across the grass.'

I had to admit it wasn't something I'd really thought about, but the more the window cleaner raved about peacocks, the more appealing it sounded. He explained that not only are peacocks beautiful, but they also make good guard animals.

'No one will get past them without you knowing about it,' he told me.

With Kelvin away for work, I had felt a little too isolated and the idea of having some noisy back-up for Ginger when it came to scaring off intruders was very appealing. But how do you get hold of a peacock?

'I can get you a couple,' the window cleaner said.

It turned out that the window cleaner knew a lot about birds in general, as the owner of one of the biggest aviaries in Britain.

'How much?' I asked, expecting him to quote me a figure in the thousands. After all, you associate peacocks with stately homes. They were bound to be expensive.

'About forty quid each,' he said.

'Really? Forty pounds?'

That sounded like a serious bargain. Imagining Kelvin's delight when he looked out across the front lawn and saw our very own peacocks, I made a snap decision.

'I'll have two.'

*　　*　　*

When Kelvin came back home from Budapest I told him what I'd done, and he seemed pretty pleased with the idea. We both liked the idea of a peacock in our garden, fanning out his magnificent tail. What a statement that would be.

We drove to meet the window cleaner at his aviary with our trailer hitched to the back of the car, ready to bring our birds home. The first surprise was that we had to go into the aviary and catch the peacocks ourselves. There were 200 peacocks in there. You've never seen anything like it. It was like something out of a game show, trying to pin two of them down – one peacock and one peahen. And then we had to get them into the trailer.

Once they were in the trailer, we thought we could relax. We'd get home, back the trailer into the yard and get the peacocks straight out into the pen Kelvin had made for them, which we'd made sure was the right size and protected from foxes. The window cleaner had explained that we would have to keep the peacocks in a pen together for at least two weeks.

'During that time,' he said, 'they'll fall in love.'

That really touched my heart.

He went on to tell us that we should speak to the peacocks when we fed them each day so they would get to know our voices, and that we needed to put the pen outside so that the peacock could see his surroundings. That way he could start scoping out the trees for one that he and the peahen could call home when they were released from their confinement – peacock first.

'But won't he fly away?' I asked.

'No. After two weeks, the male will be so besotted with the female, he won't want to leave her. He'll make them a home in the trees and she'll join him there.'

It all sounded so lovely. Kelvin and I got back into the car, excited to be setting off on a new adventure, as peacock owners and matchmakers.

We were halfway down the motorway when Kelvin suddenly said, 'Do you think they can fly out of that gap?'

'What gap?' I asked.

'The gap in the back of the trailer.'

We'd put the peacocks in a trailer made for sheep!

We got off the motorway as quickly as we could. If those peacocks escaped, they'd cause chaos, a pile-up! Thankfully, as it happened, our newly betrothed peacocks were in no hurry to get out of that trailer. No hurry at all. When we got them to the farm, they huddled right in the back of it. We had to send Marnie in to chase them out into the pen.

After that, the wait began. How would we be able to tell if the peacocks had fallen for each other? Did the peacock have a good enough view of the trees to be able to choose the right one for his love nest? I watched their progress like the producers of *Love Island* must watch the goings-on inside the villa.

At last the first two weeks were up. We opened the door to the pen and let the peacock out, while the peahen remained inside. We'd just cut the grass. It was a beautiful sunny day. The peacock walked out onto the lawn and treated us to a view of his magnificent tail. He was stunning. He seemed to be enjoying his new surroundings as he jumped onto a brick wall. He looked about him, getting ready to fly.

'He's chosen his tree,' I said to Kelvin proudly. 'Just watch him. Here he goes.'

The peacock had indeed chosen a tree. But not to set up home. Instead, he made a brief stop on a branch before setting off again. And this time he flew until he was completely out of sight.

'What the hell …? Where's he gone?'

That was not meant to happen.

When I called the window cleaner in a panic, he told me not to worry. 'He won't have gone far,' he said. 'He'll be back for his peahen. Just give him a day or two. I bet you he's really close by.'

And indeed, over the next few days, there were sightings of our peacock all over the village. But he didn't come back to the farm to see his peahen. And it didn't seem as though his lady love was missing him very much either. The moment the peacock left the pen, I'd noticed that the peahen seemed to relax. It was as if she'd been all puffed up in indignation and now she could deflate. As the days ticked by with no sign of her peacock partner, that peahen had never been happier.

When two more weeks had passed since our peacock Romeo had disappeared, we let the peahen out. The window cleaner assured us that she wouldn't go far. After spending a month in that pen, she'd consider our farm as her home. Wrong again. Within minutes of getting out, the peahen made her great escape.

We spent two days combing the woods in search of her, rattling seed to try to tempt her down from the trees. Not a sign. And if the peahen was seen around the village, we didn't hear about it. She was gone.

So that was the end of our brief venture into peacock farming. We've long since given up on them ever coming back, but just the other day, while Kelvin was away for work and I was at home with the kids, a neighbour we'd not met before knocked on the front door.

'We've lost one of our peacocks,' she said.

For a moment, I wondered whether I was about to find

out what had happened to our birds, but the peacock in the photograph she showed me was pure white. Definitely not our peacock.

The mystery of our peacocks' whereabouts remains unsolved, but I like to think that our avian Romeo and Juliet eloped to the other side of the hill and that they're living very happily in the perfect peacock tree.

CHAPTER TWENTY-FOUR

THE SHEEP OF THINGS TO COME

KELVIN

Since that day when they came to our rescue after we got stuck in the mud, Gilly and her husband Jack have quickly become great friends to me and Liz. Gilly in particular was always generous with her time and her knowledge when it came to farming matters. We knew that we wanted her to be involved in the show, but what we didn't expect was that she would be such a natural in front of the camera from the start. Even with a lens shoved right in her face, she was perfectly at ease, able to carry on talking and get on with her work as she would have done had there been no crew around.

Gilly was always busy with her own livestock, of course, but she offered to help us start our journey towards being sheep farmers. Liz and I hadn't expected to get any animals of our own for quite a while – the peacocks had given us an early lesson in the unpredictability of raising livestock – but why shouldn't we just go for it? We had the barns, we had the fields. Gilly didn't seem to think we were totally insane.

She was always completely honest about what it would entail. One of the first things I had to learn was how to get a sheep under control so that I could check it over. There's a knack to it. To begin with I was worried that I would hurt the sheep I was trying to grab, but Gilly assured me that dithering would only make things worse. It turns out that sheep are scared of everything and they can die really easily. They can literally drop dead from fright.

Gilly explained that when it came to grabbing hold of a sheep, you didn't have to be rough, but you did have to be assertive. You also had to have a plan. Once you've got hold of a sheep, you need to hang onto it, and that usually means getting it into a pen. It's easier said than done. Luckily, I like to have a process for everything. Liz was always more relaxed about the whole thing, but I knew we had to have a process for a reason. Liz and I would bicker about it, but eventually I'd convince her of the importance of a plan. If you're moving sheep, you have to work out the route beforehand and when you're putting them in a pen; it helps to have someone ready to shut the gate behind them so that they can't run straight back out the moment they realise what's going on. That's the most important bit really. Being ready to 'shut the bloody gate!' It's a lot like wrangling toddlers.

When the show aired, one of the scenes that viewers responded to most strongly was the one where I had to scrape maggots from a sheep's foot. It was a pretty disgusting five minutes, but what wasn't obvious was that it was just a short piece of footage taken from a very long day in the sheep pen.

That particular sheep had been lost for a couple of weeks and for that reason it had missed out on the regular checks Gilly gives her flock to make sure they're in good health. That's why the foot rot was so advanced. There was no way Gilly would have let things get so bad otherwise.

What you also can't tell from the screen is just how bad the smell was. Dog shit, baby shit, vomit – they all smell as sweet as roses compared to foot rot, the stench of which stays in your nose for hours after you've finished dealing with it.

I really wasn't hamming it up when I started gagging. The smell of that sheep's foot rot was like the smell of death – which I suppose is was in a way. And the sight of the maggots crawling out of the wound when we doused it with insecticide was indescribable. That was like something out of a horror film. I don't know how I didn't vomit.

After we'd cut away the worst of the foot rot and got rid of the maggots, Gilly gave the sheep a shot of antibiotics.

That sheep must have felt such relief to have those maggots gone. A couple of weeks later, the sheep was fine. And eventually so was I. Though it was touch and go for a while that afternoon.

*   *   *

With that foot rot, I'd definitely had my introduction to the ugly side of sheep farming. It wasn't something I'd ever encountered on *Emmerdale*. Every day that I shadowed Gilly I was reminded that farming's not all leaning on a five-bar gate admiring pretty little lambs as they frolic in a green field. Was I prepared to deal with something like foot rot on a regular basis? Gilly made sure I understood that the welfare of a farmer's livestock is entirely his or her responsibility.

'If one of your animals gets ill, it's down to you to fix it,' she told me. 'Those animals rely on you 100 per cent. They can't tell you what's wrong or ask for help. Their happiness and comfort is down to you and you alone. You've got to be on the lookout for problems all the time and you've got to know what to do about them.'

Until Gilly put me straight, I think I'd assumed that when a sheep got taken ill, you just called the vet and stood back. But while I was learning about the things that can go wrong with a sheep, I was also learning about the economics of farming. If I called a vet out every time an animal needed worming or treating for some minor injury – such as foot rot – I'd never make any money. To think that, back when I was persuading Liz that we should buy the farm, I'd thought it would be a less stressful way of making an income than acting!

It's hard to understand until you start farming that every animal has a price tag, and you have to make decisions accordingly. That's why I made an effort to learn how to administer vaccines to my own animals and to get over my squeamishness when it comes to foot rot. It hasn't been easy, but it's been worth it.

There were other surprises to come, such as when Gilly had me help her prepare some of her rare-breed sheep for shearing. I hadn't really considered that the sheep needed to be clean before they were sheared. I thought the wool got washed afterwards. It was eye-opening to be handed a pair of scissors and told that, before the shearer could get to work, I needed to cut away the shit that got caught up around the sheep's bums. That stuff was as hard as rock. I didn't ask if it always got like that, which is something that would come back to haunt me later on. At the time, I just did as Gilly asked me, trying not to breathe in the stink and concentrating on keeping the sheep steady so that the shit was the only thing I cut off!

I was quickly finding out that sheep farming is a messy, smelly business, but I found shadowing Gilly as she went about her work really interesting. I took it very seriously. I wanted to make sure Gilly knew that I wasn't just paying attention for the sake of good telly. I didn't switch off when the cameras stopped rolling. I apprenticed myself to Gilly wholeheartedly, humbled that she was kind enough to share her knowledge. I really wanted to learn.

My determination to be a good student served me well when Gilly took me to buy our first sheep. Normally, I would have bought the sheep at market, but there were no markets on at that time because of the Covid lockdown. I thought we'd just have to wait, but Gilly had other ideas.

Gilly had a good friend – Jean – a rare-breed farmer getting close to retirement, who was looking to downsize her herd. Perhaps I could buy some sheep from her?

As soon as I saw Jean's sheep – beautiful Cotswold sheep with thick fleeces and fluffy topknots – I knew I wanted to take them back to my farm. I'd been reading up on my sheep breeds and knew that a Cotswold's fleece could fetch good money from knitwear designers looking for the very best-quality wool.

Jean was welcoming but she had her doubts about selling the sheep she'd cared for so well to a first-timer. To her, those sheep weren't just a way of making a living. She rightly wanted to know that the person who took them over would expend as much effort on their care as she had.

I understood Jean's concern, but I didn't know what I could do to reassure her – I *was* a first-timer and I *was* inexperienced. I couldn't pretend otherwise. But Gilly stepped in to put Jean's mind at rest, telling her that she would keep a close eye on how I got on. Knowing that I had Gilly's stamp of approval helped Jean look at me as a serious buyer. I, too, made a point of stressing, 'Jean, I promise I will look after these animals.' She studied me for a moment with her shrewd eyes, then nodded in approval. We came to a deal. I would buy five ewes and five lambs.

We agreed a price, but when Jean handed me the invoice I straight away noticed that she'd undercharged me. The moment I spotted her mistake, I drew her attention to it. What the TV didn't show is that moments after I paid Jean for her sheep, her husband opened his wallet and handed me a crisp new tenner. I thought for a moment that it was to do with the fact that I'd drawn Jean's attention to the mistake on the invoice, but he explained that it was 'luck money' and that giving the buyer something back was a tradition in the farming world.

A few days after I picked up those ten sheep from Jean, she called to make sure that I was happy with my purchase and that the sheep were settling in well. I was pleased to be able to assure her that they were. Like a proud new father, I couldn't keep myself from checking on them several times a day. Liz and the kids were thrilled with them too. We had our first animals. We were on our way to being proper farmers at last.

CHAPTER TWENTY-FIVE

A COMPLETE SHEEP SHOW

LIZ

Filming our farming show really pushed us to get things up and running faster than we'd expected. I'd taken a pretty hands-off role while Gilly showed Kelvin the ropes, but once we had our own sheep, that had to change. Kelvin gave me and the kids a crash course in sheep wrangling. It was a crash course in sheep psychology too. When you've only ever seen sheep grazing quietly in a field, as was the case for me growing up, you can be forgiven for thinking they'll be easy to herd from here to there. But it turns out there's a reason why sheep farmers use really fast dogs.

Kelvin and I are normally pretty chilled out with one another – we're not given to explosive rows – but there were definitely tricky moments when we were trying to get our new sheep in and out of their pen! With Marnie still only four and Milo starting to toddle, I was used to being kept on my toes. The sheep were something else, though. They were expert escapologists. It didn't take me long to understand why Kelvin was strict about needing a plan. In the battle of man versus sheep, the sheep definitely had a few tricks up their woolly sleeves.

Jean's Cotswold sheep – now our sheep – were great beauties in the sheep world and Kelvin knew right away that he wanted to show them at the Ashbourne Show. But showing sheep isn't just a matter of displaying them in a pen. Sheep shows are more like dog shows, with competitors having to walk their sheep around a ring and getting them to stand nicely while a judge checks them over. Even when I heard this, I thought, *How hard can it be?*

During lockdown, we'd welcomed Ginger the Cavapoochon puppy into the family and we'd managed to get her house-trained. We could even occasionally persuade her to do as she was told if she thought there was a treat in the offing. Surely, training a sheep would be similar?

As with so many things, it turned out we had no idea.

Training a sheep is nothing like training a dog. Dogs are generally motivated by wanting to please their owners. Sheep don't give a stuff what you want. Dogs can be pretty intelligent. Sheep, as a general rule, are not.

Still, Gilly said it could be done and we trusted her view. She told us that we should aim for fifteen to twenty minutes of training each day, first getting our sheep used to wearing a head collar, then motivating them to walk with the promise of sheep pellets. At least, like dogs, they were interested in food.

Kelvin chose the three best-looking sheep for our Fletcher family training programme: a ewe and two lambs – one male and one female. They certainly looked the part with their impressive topknots (a feature of the Cotswold breed), but our early attempts at training did not go well. If we managed to get our sheep into their halters, they would immediately fling themselves onto the floor, playing dead. If the worst thing about sheep is foot rot, the second is their absolute stupidity. Kelvin would creep into the pen, moving slowly, doing his best to keep them calm, but one would always end up trying to vault the fence. And once one sheep panics, they all panic. It's mayhem.

'Why do farmers do this?' I asked Kelvin as another training session ended in a stampede. 'How do they find the time?'

It wasn't as though there were big financial rewards for training a prize-winning sheep. A slight uptick in the value of a prize-winning tup – a breeding male – perhaps, but otherwise?

'It's all about pride,' Kelvin told me. 'It's about showing our fellow farmers what we can do and how well cared for our animals are. I want to do us and Jean proud.'

So it was back into the sheep pen, day after day, hoping that at least one of the sheep would 'get it' and let us parade him or her on a lead. I'm sure Ginger the dog must have been laughing from the other side of the fence.

KELVIN

Our adventures in sheep training marked the first time I had come to see a competitive side to Gilly. Up until then, I had been certain that if I asked her a sheep-related question, she would give me an entirely honest answer. I knew that she would never neglect to warn me about something that could impact the welfare of an animal in my care. She was too professional a farmer and much too good a person for that. But when it came to a sheep *show*, all bets were off.

After days of training my sheep and not really seeing any progress, I began to wonder if there was something Gilly wasn't telling me. She was going to be taking some of her own sheep to the same show and seemed very relaxed about the competition ahead. Very relaxed indeed. Meanwhile, we still had three sheep that played dead at the slightest provocation, no matter how many of their favourite sheep nuts we offered them. In fact, the more sheep nuts we gave those sheep, the more likely they were to act as if we were trying to kill them every time we got the halter out. The familiarity of the training routine didn't seem to be making it any easier. If we'd spent as much time training Ginger the dog, we could have coached her to win the glitterball on *Strictly*.

Eventually, we managed to persuade the sheep to at least stay upright with their halters on. That would have to be good enough.

'If we just make them look pretty, that will go a long way, won't it?' said Liz.

At least we knew that our sheep were from prize-winning stock and they could look the part.

We waited until the day before the competition to give our sheep a bath. There didn't seem to be much point in starting the clean-up operation earlier. They were sheep. They lived in a field and spent all day eating and pooing. If we groomed them too early, we'd only end up doing it again. Big mistake …

Getting those sheep ready for the show was not going to be like giving Ginger a bath and a blow-dry. I should have learned my lesson when Gilly had invited me to spend the day watching her sheep being sheared and she'd asked me to help speed up the process by cutting away some of the shit that dangled from her sheep's backsides. I'd been astonished then at how difficult it was to get that stuff off. It was rock hard. It didn't brush away. It had to be cut out of the fleece. I don't know why I thought it would be any different with my own sheep.

One of the things Gilly hadn't warned me about was that I should have started getting the animals show ready weeks before the big day. What I didn't know was that some of the farmers we'd be up against would have been keeping the sheep they were going to show separate from the rest of their flocks for months. Those show sheep spent their days in separate paddocks and slept in pristine sheds. They were barely allowed to see mud and their bottoms were checked for dangling bits every day.

But we'd let our sheep be sheep, and as a result we were left with a problem that soap and water would not fix. It turns out that you have to approach preparing sheep for a show like looking after a classic car. You can't just put it through the car wash once a year and expect it to look like you've just driven it off the forecourt. The muck dangling from our sheep's bottoms put me in mind of concrete that's been sprayed where it shouldn't have. We were going to be up all night chipping it off.

LIZ

It was hard work getting those sheep tidied up. Unfortunately, the skills I'd picked up at my hair and beauty course in Rochdale weren't much help. But having done our best, on the day of the show itself we piled the sheep into the trailer and set off with a sense of cautious optimism.

That optimism took a wobble when we saw the competition at the showground. Cotswold sheep are naturally shaggy and they're never going to look as clean and fluffy as the winning bichon frise at Crufts, but there were certainly other breeds of sheep at the show that wouldn't have looked out of place on a catwalk. They were immaculate. Some of them even had polished hooves. The best we could hope for was that our sheep would suddenly transform themselves into models of good behaviour in the ring.

Kelvin went first, taking the ewe. His sheep stayed on its feet, but he couldn't persuade it to walk nicely. She dug her hooves in and stayed put. Kelvin wasn't placed. Me next, with Daniel, the male lamb. The minute we were in the ring, he executed a perfect death flop, which would have been great were he an actor pretending to have been shot in a war film and not our best hope for a sheep-show rosette.

Two down. The whole family's hopes were pinned on Marnie's lamb.

But Marnie, overwhelmed by all the people crowding the show ring told us she didn't want to do it. Gently, we reminded her how much she'd enjoyed training her lamb, which she'd called Rosie. Didn't she want to show everyone how clever at sheep training she was? Just in time for her class, Marnie put on the white coat embroidered with her name and took hold of her lamb's halter lead.

The lamb Marnie took into the ring that afternoon was easily as big as she was. It could have knocked her over or refused to move. But something magical happened as they stepped out together. Marnie rested her hand on the lamb's back and it seemed to grow calm at her touch.

We're proud of Marnie every day, but that afternoon I thought my heart would burst at the sight of our girl taking her moment in the show ring in her stride. She persuaded her lamb to stand perfectly still while the judge made her rounds. Then, when she was asked to take the lamb for a walk, she had it skipping around the ring as though it was a little white puppy.

Marnie's lamb came third in its class. When the judge handed Marnie a rosette, she beamed. We knew then that the next time a sheep show was in the calendar, far from having to persuade her to go into that ring, we wouldn't be able to stop her.

KELVIN

With the exception of Marnie's rosette, we came away from our first sheep show empty-handed but full of pride that we'd gone out there and done our best. We'd been on a sharp learning curve and, having watched everything closely, we'd picked up some tricks to try next time. I noticed, for example, that Gilly didn't exactly 'walk' her sheep, but jogged along with them, giving a sense of energy and also playing to the sheep's natural tendency to want to run away.

It had been hard work, but we knew we would take our animals to shows again. There was the Great Yorkshire Show, the Staffordshire Show, the Royal Agricultural Show …

As I told Liz and Marnie, 'One day we'll have a champion sheep on our hands.'

A few days after the show, we received a letter from Jean, the farmer who had sold us our beautiful Cotswolds. Marnie's win had made the local paper and, thinking we might not have seen the report, Jean had sent us the cuttings. It was such a lovely gesture and one that made us feel a little closer to being considered part of the local farming community. We were glad that Jean could see that we were taking proper care of her lambs and that, in

Marnie, a new generation was getting ready to take on the world of rare breeds.

*   *   *

One thing we've discovered since we moved to the farm is that Marnie is a natural when it comes to caring for animals. Dogs, sheep, cows, pigs. She loves them all and they seem to love her in return. What's more, she's not put off by the aspects of farming that you'd think a child would find hard to handle.

When we first moved to the farm, Gilly called to ask if we would be interested in watching a cow giving birth. One of her cows was about to have a calf. I wasn't sure I could handle it. Too squeamish. But Marnie insisted on going along – which meant that Liz had to go too – and she was right there at the side of the pen, watching every minute, not even blinking when the vet had his arm right inside the cow. Even the vet's son, who must have seen this kind of thing a dozen times, looked queasy. Marnie was fascinated.

More recently, I was at Leek Market with Gilly, tagging along when she went to buy some cows. A farmer – a real old-timer – stopped me and said he'd seen our farming show. 'Your daughter,' he said. 'She's got it.'

He told me he'd seen lots of children who'd grown up on farms, but it was rare that they had Marnie's confidence

with livestock. I was so proud to hear it. That the acknow-
ledgement was coming from a farmer meant even more.

I wonder if Milo will follow in his big sister's footsteps.
We have to be careful while he's still so young, as he has
little to no understanding of danger, but he can already
round up the sheep. Having watched Marnie in action, he
knows the arm movements he needs to make to bat them
into corners without scaring them. He enjoys being given
little tasks. But for now, he enjoys being half-naked in the
pigs' trough even more. Liz says he takes after me ...

CHAPTER TWENTY-SIX

THE CIRCLE OF LIFE

KELVIN

I'm sure that many people watching our farm show thought to themselves that we weren't doing it for real. With so much scripted reality TV on our screens, I can understand why viewers might think that Liz and I were spending most of our time in a hair and make-up caravan somewhere out of sight, coming out only to pose with something cute. Believe me, we were and are *really* farming. Not that it's all mud and muck. There's a hell of a lot of paperwork too.

One thing I definitely wasn't prepared for was the paperwork involved in keeping livestock. When you buy sheep, for example, you don't just hand over the cash and load your new animals into the back of a van. There are loads of forms to fill in: CPH (county parish holding) forms, movement forms, vet's forms, forms for the accountant ...

This paperwork gets checked all the time. You can't afford to get it wrong.

Once again, the first time I had to do it, Gilly came to the rescue, holding my hand through every step until I was confident I knew what I was doing.

Unfortunately, Daniel, the lamb that Liz showed at Ashbourne, died unexpectedly a few months later. We were shocked to find him dead one morning. Though he'd looked a bit sad for a couple of days, there was no reason for us to think he was unwell. I rang Gilly, who commiserated but reminded me that no matter how well you care for them, sheep often die without any clear reason.

'If you've got livestock, you've got dead stock,' she said.

Daniel was the first animal to die on the farm under our care. Dealing with dead livestock was another lesson we needed to learn. Daniel's death had to be noted in an official book and his body disposed of in the official way. Gilly gave me the number of a local company who would come to collect him.

Losing Daniel was tough. Because he'd been part of our show team, we felt as though we'd got to know him, but the fact was that the intention had never been to keep our sheep as pets. We were supposed to be making money, both from the sheep's wool and from their meat. That meant learning how to say 'goodbye'.

Liz and I grew up in the city, and certainly as kids we never really thought about how the meat we ate for Sunday lunch ended up on the table. It was easy not to make the connection between the sheep in the fields and the lamb on our plates. That was all about to change.

Sending one of our male lambs off for slaughter was one of the hardest things I've ever done. I could hardly bear to look at him as I loaded him into the trailer for the trip to the abattoir; I felt such enormous guilt for choosing to end his life. But I remembered all the conversations I'd had with Gilly about this most difficult part in the life cycle of a farm animal. There was no room for sentimentality. I knew that if I couldn't do this then I shouldn't be a farmer.

We chose to show the moment when we took the lamb to the abattoir on our docuseries because we believe it's important to be honest about the farming life and the role farmers play in the food chain. I think that if we're going to eat meat at all, then we need to be aware of where it comes from and also how animals are treated on our farms.

I've come a long way from that day when Gilly showed me how to deal with a sheep that's got foot rot and I nearly threw up from the combination of the smell and the sight of the maggots. I know now that there's no place in livestock farming for someone who isn't prepared to get their hands (and everything else) dirty. At the end of the day, you can never forget that farm animals are sentient beings. You're dictating their destiny and with that comes a responsibility to make sure that they have the best possible life. It's hard to send an animal off to the butcher, but knowing that you have done everything you could to make sure that it was happy and comfortable throughout its life – from the very beginning – makes it a little easier.

My brief experience of farming has already convinced me that we need to recognise the dangers of over-farming and to reconnect with nature. I understand why other farmers may disagree. Overheads are always rising. The price of fuel and farming chemicals has risen four-fold. That cuts into profits. It's easy to understand in these circumstances why there's a move to rearing animals in sheds, where they grow more quickly.

But there has to be a middle way. Prices are rising, yet way too much of the food we see in the supermarkets ends up being thrown away. We've moved away from eating seasonally, expecting strawberries all year round. Farmers are underpaid for what they produce. If we bought in season, bought less and thus required the farming community to produce less, we could make sure that the food we eat is better and more fairly priced. Just as importantly, we could make sure that the animals we raise have better, happier lives. We could move towards a more sustainable, ethical way of living. That's what we hope to achieve.

Later that day, I brought part of the carcass of our first lamb back to the farm and cooked a small part of it for supper. I have to admit that Liz and I found it hard to tuck into the lamb that we knew had been walking around our fields days earlier, but Marnie, our future farmer, pronounced it delicious. It was a surreal moment but a very important one.

CHAPTER TWENTY-SEVEN

GARLIC BRED

KELVIN

At the end of the summer, we celebrated the end of filming our first series of the farm show with a party in one of our fields, to which we asked all our new neighbours. When we went into our local village to post the invitations, we had no idea we'd get such a lovely, warm response. A group of villagers asked us to meet them at the pub and told us they'd be delighted to help turn our farm-warming party into a real community event – much needed after all the Covid lockdowns. Before long, we had offers of bunting, tents, rounders kit and a rope for a tug-of-war. A team from the local Women's Institute offered to take care of the catering. When Liz turned up to make sandwiches, they made her an honorary member.

Despite a small problem with the tea stand – we had hoped to use water from our well, but we weren't sure it would be clean enough – the party went brilliantly. More than a hundred people turned up to picnic, play rounders and catch up with friends old and new. It was great to be able to extend our hospitality to our neighbours and feel so welcomed in return. It was the best way to mark our first half-year on the farm.

But as Gilly and Jack reminded us over a glass of wine at around the same time, we were only just getting started. We couldn't call ourselves farmers yet. We hadn't even done all the seasons.

'You've got to prove you can do it,' agreed Marnie, sounding old beyond her years, as she sat between our neighbours at the kitchen table.

'She knows what she's talking about,' said Gilly.

*   *   *

As summer became autumn, a whole new farming season was about to begin. Now that we were used to caring for the Cotswold sheep we'd bought from Jean, it was time to think about expanding our flock – both by buying some more sheep and by breeding from those we already had.

Gilly to the rescue again. She helped us to buy a small flock of mules. The next stage was to buy a couple of rams – one Cotswold, to breed with our Cotswold ewes, and one Texel for the mules. Gilly invited me to join her at an auction when she went to buy some rams of her own.

But how do you choose a ram?

'You'll know instinctively,' Gilly assured me. 'If it looks like a good animal, it probably is.'

So far, so simple.

'You'll need to have a feel of their bums.'

'Right.'

'To get a sense of their real size. See if they're corned up.'

What did that mean? Gilly explained that ahead of auction, some farmers are guilty of doing something to their sheep called 'corning them up', which means feeding the sheep up so that they look bigger than they really are. You think you're buying a great big animal. Two weeks later, it's back to normal size. At the same time, you don't want your ram to be too big – certainly not round the head – in case the lambs he sires are too big to be born without intervention. Certain breeds are easier when it comes to lambing. We were too inexperienced to pick a difficult breed.

'You've also got to get a sense of the ram's personality,' Gilly told me. 'What kind of temperament have they got?'

One ram that I had liked the look of from a distance was wild in the pen, continually jumping at the fence. A sheep's instinct is to get away from danger, but this particular sheep took that to the extreme.

'And don't forget to check its balls,' Gilly concluded.

Perhaps that wild-looking ram had good reason to be so jumpy after all.

When I walked around with Gilly, I found there was a 50-per-cent crossover between the rams I was interested in and the ones she thought might be a good buy.

The rams would be sold by auction. As with so many things, I looked to Gilly to get a sense of what I should expect to bid. I knew that recently Gilly had paid just over £800 for a Texel. It was slightly more than I wanted to shell out. I hoped I'd be able to get the ram I wanted for less. Perhaps I'd be lucky and no one else would be after the same one.

LIZ

Kelvin was insistent that the children and I join him at the auction that day. I agreed with him that it would be great fun for Marnie and Milo to see all the animals up close, but I was much less excited. We'd recently learned that I was pregnant. We hadn't told anyone else yet. Although we were planning to have another baby, the news had come sooner than expected. I was supposed to be going into hospital to have my tonsils out. At the pre-op appointment, the doctor asked if I could be pregnant. I didn't think I was, but I took a test anyway and there was the positive result. It showed up within seconds. I didn't have to wait.

Kelvin wasn't at home that day. He was away racing. When he came home and I told him the news, he was over the moon. Unfortunately, since finding out we were having another baby, I'd been suffering from terrible sickness – far worse than I'd ever experienced with Marnie or Milo. For that reason, the very last thing I felt like doing was standing around at a livestock market. I knew how important it was that we find the right tup, but I had hoped Kelvin would be happy to do that job by himself. I just wanted to get through the day as quickly as possible so that I could get back to feeling sick in the comfort of my own home.

The kids were having a whale of a time, running around, eating ice cream. We'd thought they wouldn't be welcome at the auction, but we were pleasantly surprised by how relaxed the farmers were about having children around. So long as they didn't touch anything they shouldn't, everything was OK. It was rough and ready, but this was what we'd wanted for our children – a life full of adventure where they weren't always being told 'no'. They loved being around the animals and they especially loved hanging out with Gilly and Jack.

Kelvin was busy making new farmer friends, happily accepting the tots of whisky they offered from their flasks.

We felt welcomed into the community, though perhaps there was an element of the old-timers wanting to see how we would navigate this new hurdle – buying a ram at auction – for the first time.

Kelvin had earmarked the ram we were going to bid for. He'd checked it out properly and had the farmer who was selling it promise that it hadn't been 'corned up' and thus wouldn't end up looking scrawny and deflated in a couple of weeks.

I couldn't believe my rotten luck when I saw that the Texel tup we wanted was very last on the list to be auctioned that day. I was already more than ready to go home. For Kelvin, though, having to wait through all those other auctions was a good thing. It gave us a chance to watch and learn exactly how bids were placed.

'We don't want to look like amateurs,' he said.

The afternoon dragged on, with Kelvin paying close attention to every auction. They happened fast. The sheep would be brought into the pen and the auctioneer would start talking at a million miles an hour, taking bids from all around before bringing the hammer down. It happened so quickly, I couldn't ever keep up with where the bids were coming from and was always surprised to see who'd won.

At last, the ram we wanted was brought out. I could see why Kelvin wanted him. He looked good. He was sturdy and calm. Kelvin had got himself into a good position to bid, standing right at the side of the pen. I stood behind him and waited for the action to begin.

The auctioneer started talking at high speed. The bidding started at £500. Seconds later, it was £520, then £530, then £540. Yet Kelvin hadn't made his move. I couldn't understand it. He seemed to be in his own little world, just standing there staring into the pen, letting the tup he said we wanted be sold from underneath him.

'Kelvin?' I tried to get his attention, but he wouldn't turn around. What was he thinking? 'Kelvin!' Was I going to have to take matters into my own hands?

KELVIN

All afternoon, I'd watched the sheep auctions as carefully as I could, trying to get a sense of how it all worked. It wasn't easy to follow the action to begin with. The auctioneers spoke so quickly and the farmers who bid on each lot did so subtly, with just a small nod or hand gesture to indicate they were happy to go to the next level. No one was waving their arms in the air like you see in auctions on the telly. I modelled my bidding style on a farmer opposite, who was so cool I hardly knew he'd been bidding at all until I saw him being handed his luck money.

When the tup we wanted was brought into the ring, I quickly took the measure of the other farmers leaning on the rail. Did any of them seem to be getting ready to bid against me? I wasn't sure.

The bidding began. I gave a nod and the auctioneer acknowledged me, then turned to the cool farmer across the way. To him, to me, to him, to me. When the bidding reached £540, my rival shook his head to indicate that the price was too high for him. Great. The tup was mine and for a lot less than I'd hoped to pay for it. But ...

'Five-sixty.' The auctioneer continued taking the price up.

'What?'

I nodded for £580.

'Six hundred ...'

I went for £610.

'Six-thirty ...'

Who the bloody hell was bidding against me? I was vaguely aware that someone behind me kept putting their hand up, but I dared not look round to see my competition. Soon the bidding had reached £650. It was the limit I'd set for myself. Reluctantly, I bowed out. I'd lost my tup. I turned to Liz to express my disappointment, but she was jumping up and down.

'We got it, Kelvin, we got it.'

'Got what?'

'We got our tup.'

Oh no. The mystery bidder was my own wife. She'd been bidding against me all the way from £540 and she was so pleased with herself for having won.

'Liz ...' I sighed. 'You bid against me. You weren't supposed to get involved.'

'But you were just standing there. I had to do something. I thought you weren't bidding.'

'I *was* bidding,' I said. 'For heaven's sake ...'

Liz shrugged and gave me her winning smile.

'Oh well. At least we can go home now,' she said.

Eventually I saw the funny side. Though we'd paid more than we needed to, we hadn't actually gone over budget and the tup I'd chosen turned out to be every bit as good as I'd hoped. After two weeks at our place, he looked as big and strong as ever. He definitely hadn't been corned up ahead of the sale. He had a good temperament too. Perhaps it was beginner's luck, but we were happy with our purchase. And Liz and I had learned a very good lesson about agreeing which one of us was going to bid before an auction started.

*   *   *

In farming, as in acting and dancing, timing is everything. When it comes to breeding sheep, the first thing you have to take into consideration is when you want your lambs to be born. And that means paying close attention to when you let the sheep mate. You also have to pay close attention to which of your sheep have tupped. That's where the coloured marks you see on the backs of sheep in the fields come in.

Gilly had told me about 'raddle', the colourful paste that would tell us which ewes the tup had served. It goes on his chest and ends up on the ewe's back when they mate. What I didn't know was that raddle doesn't come as a paste, which is what I expected. It comes as a powder that needs to be mixed to the right consistency with fat or oil.

'Lard, butter – anything like that will do,' Gilly told me.

Great. I went straight to our kitchen and opened the fridge. Liz wouldn't mind me helping myself. Except that we didn't have any lard or butter in the house. We needed to go shopping. We didn't even have any vegetable oil.

'Liz?'

Liz came in as I was going through the cupboards again in search of something, anything, I could mix with the powder. I didn't believe we were all out of everything.

'I need some oil. For the tup.' I wanted to get the Texel out with the ewes as soon as I could.

'What kind of oil?'

Understandably, Liz did not want to sacrifice any of her expensive body lotion for the project. She joined me in going through the kitchen cupboard.

'How about this?' she asked me, waving a bottle of garlic-flavoured olive oil.

'Garlic?'

It was the only thing we had. It certainly made the process of mixing the raddle up interesting. We just had to hope that it wouldn't put the ewes off in the way garlic always seemed to put the ladies off back in my clubbing days.

Fortunately, the ewes didn't seem to mind a tup who smelled like garlic bread. Neither did the ewes on the farm next door. One morning shortly after we'd started the tupping process, I went down to the field to see that our garlic-infused tup had escaped. I searched high and low but couldn't see him anywhere. I did the school run, planning to look for him again when I got back. Once I'd dropped the kids off, I drove every inch of our perimeter fences. I walked every inch of the woods. The tup was nowhere to be seen. That's when our neighbour appeared on his quad bike. He did not look happy.

'Morning,' I said.

Our neighbour didn't bother with the usual greetings but cut straight to the chase.

'Your tup is in my shed,' he said.

'What's he doing in there?'

'I found him with our ewes He's served two of them already.'

'Has he? Good lad!'

My joke fell flat, and it quickly became clear that it was not a laughing matter.

'You better come and get him.'

I hooked up the sheep trailer to the Gator and drove over to the farm next door. My neighbour led me to the shed where my tup had been cornered. My neighbour had not recovered his sense of humour and the tup did not look too happy either. There was nothing sheeplike about his demeanour that morning; I could tell that he was ready to take out his frustration on someone and that someone was most likely going to be me.

My neighbour and his family stood back and watched me try to persuade my 110kg tup into my trailer single-handedly. It was a tense moment. Eventually, I had my tup loaded up and was ready to go, with my neighbour muttering darkly about compensation. I got out of there as quickly as I could.

'What does he want compensation for?' I asked Gilly later that day. 'At worst, he's going to get some free lambs out of it.'

'Yes, but he doesn't want *those* lambs,' Gilly told me.

The only good news was that our tup had only managed to serve two ewes – as evidenced by the strong smell of garlic raddle – before my neighbour saw what was going on! We've kept a much closer eye on that tup ever since.

CHAPTER TWENTY-EIGHT

BACON AND LEGS

LIZ

Having got the hang of working with sheep, we decided it was time to diversify. Part of the land we bought with the farm is a forested area on a steep hillside. The previous owner of the farm planted the trees as part of an ecological programme and it looked great. Unfortunately, it was no good for livestock. Except perhaps pigs. That forest, which contained oak trees, was a potential piggy paradise.

We bought three weaners – pigs around two months old. The plan was to keep them until they were six or seven months old before selling them on for pork. We chose Oxford Sandy and Black pigs. Their pork tastes amazing. They don't grow as quickly or get as big as some other breeds, but they can be kept outdoors. Our little forest would be the perfect home for them.

To begin with, the weaners had to live in a pen, which is how we quickly learned that pigs are expert escapologists. Kelvin had built the pig pens knowing that pigs were strong, but we really underestimated what they could do.

On the pigs' very first night in the farm, we were sitting on the sofa, finally relaxing after a busy day, when something caught my eye.

'Kelvin,' I nudged him, 'is that our pigs?'

It was! They were trotting right past the lounge window. So much for chilling out in front of the telly. It was all hands on deck again to get those pigs back in their pen. And to think I'd thought sheep were hard to handle!

Watching the footage of the escape back on the CCTV, we were astonished by the cunning and sense of mischief our new pigs had. The biggest of the three was definitely the ring leader, looking for the weak points in the pen before encouraging the other two to follow her to the great outdoors.

'We're never going to be able to keep them in,' I despaired that first night.

Luckily, pigs are massively motivated by food so it's relatively easy to persuade them to come back to where you want them by bribing them with something to eat. They're always hungry. Always wanting to be fed. Once they've got their heads in a feed bucket, there's no way you're getting them out until every last morsel is gone.

Occasionally, when we're in the pen feeding the pigs or mucking out, they'll nip at our ankles, which is a bit unnerving since Gilly told us the horror story of an uncle of hers, who died of a heart attack while feeding his pigs. They made short work of his corpse. I have to say I looked at our three differently after hearing that.

At the same time, they love to be stroked and patted. They're very inquisitive, love visitors and are always happy to roll over for a tummy rub. They have such cheek and charm. While their favourite thing is to get into some horse manure, there's no doubt that they are lovable creatures.

It didn't take long for us to fall in love with our first three pigs. When the moment came to send them off for slaughter, we somehow conspired to forget. Instead, we decided to keep them and see if we could breed off them instead. A year later, they're still having the time of their lives as Kelvin builds ever bigger pens and outdoor runs to accommodate the fact that our cute little weaners are now very definitely porkers.

We didn't start out as animal people, but we were going that way. There was a great deal of stress and responsibility involved in raising the animals, but there was also the opportunity to really make something of the farm. Unlike in our acting careers, on the farm our efforts were rewarded with direct results. The idea that we could commercialise the farm was exciting. It was good to feel secure, although any farmer would tell you that farming is far from a licence to print money.

After we bought out pigs, we looked at other livestock that would be suited to our land and might help us turn a profit. Our first efforts at making money from our Cotswold sheep's wool had not gone quite as well as we hoped, thanks to the horror that is sheep's muck – by not understanding how clean sheep needed to be before they were sheared, we'd lost money, paying the shearer more than we got for the dirty fleeces – but perhaps there were other animals we could raise for wool. Alpacas? The parents of one of Kelvin's motor-racing friends were kind enough to make us a gift of three.

Sheep farmers have been turning to alpacas for lots of reasons. Not only can they provide excellent wool, but they also make very good guard animals. They're often put in fields with lambs as they can deter foxes hoping for a nice dinner. They're low maintenance. They're very happy so long as they have somewhere to graze. They're beautiful creatures – so classy, with their long legs and silky eyelashes. Except when they're mating, that is!

As anyone who has seen the episode of our show where we feature the alpacas will know, the breeding of alpacas is quite an undertaking. It's not just a matter of hit and miss as it is with the sheep. When they see a female, male alpacas make an unearthly-sounding mating call. It's called 'orgling'. If a female alpaca is already pregnant, she simply ignores the invitation, and if the male persists, she'll spit at him. This reaction, which is called a 'spit off', is a reliable way of telling how a pregnancy is progressing. If she's not pregnant, the female will actually ovulate to the orgling sound and get down on her knees to allow mating to take place.

Female alpacas simply won't mate without hearing that orgling. Because it literally makes them ovulate, if you're going to artificially inseminate an alpaca, you have to imitate the orgling too. Some farmers might be brave enough to try orgling themselves, but luckily you can play a recording. Kelvin and I have learned many things since we moved to the farm, but I don't think we'll be trying orgling any time soon.

Alpacas are such gentle creatures. There are farms that offer alpaca yoga or alpaca picnics. One day I'd like to open up our farm to visits from children who might benefit from the calming influence of an alpaca's majestic presence.

I fell instantly in love with our three alpacas when we brought them home. When they're out in the field they stick closely together, and whenever I see them I always imagine them saying to one another, 'I don't know why we're stuck in here with these filthy sheep. We're not live-stock, you know. We've got class.'

CHAPTER TWENTY-NINE
IT'S TWINS!

LIZ

With a full farm – sheep, pigs and alpacas – we hoped we could look forward to a quiet winter, but of course the best-laid plans ...

Throughout the filming of our show – particularly when we were recording anything to do with breeding our animals – I'd batted away the questions and comments about me and Kelvin being next with a swift 'no way'. Marnie had only just started school. Milo was still under three. I already had plenty of work to be getting on with. But as was public knowledge by the time the show aired, while I was busy saying 'no', I was actually already pregnant.

Just before Christmas it was time for my twelve-week scan. This twelve-week scan wasn't a big deal. The first time I went for a scan, when I was pregnant with Marnie,

Kelvin and I cleared the whole day, expecting to be over-whelmed by the experience of meeting our baby on screen. And we were. The same applied to the twelve-week scan for Milo. Seeing him on screen for the first time took our breath away. This time we were excited but taking it in our stride, like we were going to the post office! Every parent who has been through three pregnancies knows the score. Not to mention the fact that there was so much to do before we went to the hospital. The animals needed to be checked, we had to do a school run, the builders working on the cottage needed to discuss how work was progress-ing … There was barely time to brush our hair before we had to leave for Wythenshawe Hospital.

One of the trickiest things about being married to some-one famous is that even during our most private moments together he might be recognised. That was what I dreaded whenever we had to go to the hospital. I remember after Milo was born, I worried about being seen and it seemed harder to have my little bit of privacy.

Fortunately, that morning at the hospital, we were taken into the scanning room before anyone spotted we were there. After the chaos of the morning at the farm, it was lovely to be in such a calm, dimly lit space. It was so peace-ful and quiet.

As she moved the wand over my belly, the sonographer asked me, 'What baby is this?'

'Number three,' I said.

I should have guessed at once from the look on her face that she was finding something funny. 'I've got news for you,' she told me.

In that moment, everything made sense. How fast and strong the pregnancy test result had been. How sick I'd felt.

'Are there two in there?' I asked.

'Yes,' the sonographer nodded. 'The good news is, you've not sworn.'

I think I was about to …

There were twins on my side of the family – Nana was a non-identical twin – but all the same the news came as a surprise. How was I ever going to carry twins? As the news sank in, I started crying and couldn't stop.

Meanwhile, Kelvin was jumping with joy. 'Twins! It's a miracle!'

As far as Kelvin was concerned, there was no better feeling than being told you're having not just one baby, but two.

'It's better than winning ten billion pounds,' he said.

But I was in shock. My experience of pregnancy with Marnie and then Milo had been mixed. The memory of having to go into surgery straight after giving birth to Marnie still had the power to stop me in my tracks, as did the memory of having the baby blues. I knew just how much could go wrong, so the thought of carrying those two babies was daunting.

I was reassured that the twins would be delivered by caesarean, but after the birth? The idea of having four children … it was scary. We were already so busy. Rarely a day went by when our plans weren't upended by something to do with the children and animals we already had. And I'd been so sick throughout the pregnancy so far. Was I ever going to feel any better?

The staff at the hospital understood that such momentous news might bring mixed feelings and a million questions. We were ushered into a counselling room, where someone talked us through the risks. Though the risks were far from insignificant, Kelvin was euphoric.

'It's going to be great, Liz,' he promised me. 'This is going to be the best thing that ever happened to us.'

I sat there in silence, still stunned.

Because we needed to hear from the health experts all about the extra risks that come with having twins, we were at the hospital for longer than we expected that day. We drove home, picking the kids up from school on the way. We told them the news as soon as we got back to the farm. Marnie was as excited as her father was.

I remained low-key on the subject, but by the time we went to bed that evening the joy Kelvin had felt so instantly and fully when the twins appeared on the scan had started to come over me too. Marnie's and Milo's delighted reactions had helped. They couldn't wait to have two new baby siblings.

Of course we were having twins. Of course! Ever since I'd met Kelvin, I'd known that he didn't do anything by halves, and now, as his wife, it seemed neither could I.

*   *   *

That Christmas, our first Christmas on the farm, was a really joyful one. After the disappointment of 2020, when Christmas was all but cancelled for so many people up and down the United Kingdom, it was wonderful to be able to celebrate with family and friends again. We invited the whole family and Jack and Gilly, who had become like family to us. Unfortunately, we weren't able to cook Christmas lunch using produce from our own farm, but we did have some of Gilly's lamb, which was delicious.

Christmas Day was quite a minimal affair. There were plenty of toys for the children, but for us adults it was just nice to be together again after all the lockdowns of 2020 and 2021.

For my brothers, who were visiting the farm for the first time, it was a real experience. Of course, farmers don't really get to have Christmas off. There was still plenty of work to be done about the place. Kelvin quickly co-opted my elder brother, who is a surgeon, to help him give the pregnant ewes their Heptavac boosters, to make sure their lambs were born with immunity to all sorts of conditions. I don't think the sheep could believe their luck. My brother did the job really neatly and delicately, taking every bit as much care with the sheep as he always did with the humans in his care.

We had a wonderful few days surrounded by all the people we loved. By the time the New Year rolled around, the farm felt more like home than ever and we were ready for whatever 2022 had to bring.

CHAPTER THIRTY

LATE NIGHTS AND EARLY STARTS

KELVIN

That period post-Christmas might be a quiet time for most people, but life on the farm was as busy as ever – after all, we were preparing for a lot of new arrivals! At Liz's twenty-week scan we had a choice to make: should we find out the sex of our twins or not? I wanted it to be a surprise, but Liz was keen to know, so we agreed to find out and were made up when we were told that we were expecting two boys.

Finding out what to expect from our first lambing season was a little more complicated. Not every farmer gets their ewes scanned. It's just really good practice to do so. The biggest reason is that once you scan them you will know how many you're getting: some ewes can carry just one lamb, but others can carry three. It's important to know which is which; you don't want to overfeed a ewe carrying a single lamb, in case the lambs becomes so big she can't easily lamb. Likewise, you don't want to underfeed a ewe carrying three, in case one of her lambs becomes dangerously underweight.

The best time to do the scanning is seventy-five to ninety days after the ewes have been tupped. However, when it came to my own, and after trying to organise a scanner, it had already been 105 days, so we'd slightly missed that window. Instead, I brought them in and had a feel of each. I made careful observations about each one – the feel, the size, the shape; anything I thought was useful – and jotted them down in my notebook.

Liz wasn't twiddling her thumbs either, with the children to look after, the cottage to finish renovating and her voiceover work. One good thing to come out of the pandemic for our family was that it changed the way voiceovers were recorded forever. There was no need for Liz to rush down to London for the day any more, now that she could record everything online in our own home studio. It made a big difference not to have to do all the travelling.

In January, our farm show aired for the first time and that kept us busy with press. Meanwhile, I was still on the lookout for my next acting job when Olly rang and asked if I'd like to go to an audition at the National Theatre.

'Yeah, right,' I said. It sounded even more unlikely than when he'd called to tell me I was wanted on *Strictly*. The

National Theatre? Home of serious acting? I remember thinking at the time how life can take the most unexpected and unplanned turns. I mean, I was going from filming a docuseries set on our new farm to being called up to London to audition for a play. You just had to laugh. But still, I tried to embrace this moment because it's exactly these kinds of opportunities that really stand out. A nice reminder that when you least expect it, something always comes through. And Olly was very serious. The audition was for a part in a comedy set during the Second World War, called *Jack Absolute Flies Again*.

'It's a big opportunity, Kelvin,' Olly said. 'And they've asked to see you. And I think you'd be the perfect fit for it.'

Liz was by my side, like she has been so many times before. After my call, I turned to her, a little speechless.

'But I can't believe it's the National …'

'So what?' she said in her usual calm way.

It was my first in-person audition since Covid had hit in March 2020. I came out of the room buzzing. Stepping into that audition, I'd felt like a kid again. I enjoyed dissecting the character and trying the same scene out in a dozen different ways – softer, faster, with more assertive body language. I was really put through my paces. I lost myself in the moment and was reminded of what really mattered to me. The chance to act. If nothing came of the audition, I was glad that I'd had the opportunity to show what I could do again.

You can never really tell what the outcome of an audition will be. You can walk out thinking it went well only to hear nothing. Then you can come out of an audition thinking you performed really poorly and get the part. It's always an amazing feeling when you do get that part.

When I heard that I'd been cast in *Jack Absolute Flies Again*, it reaffirmed all the choices I'd made. I'd been given the part of Dudley Scunthorpe, a Lancastrian mechanic. Dudley was described as 'handsome and athletic'.

'You'll have to do a lot of acting then,' Liz joked.

The script, by Richard Bean, was incredibly funny, heart-warming and slightly risqué. It made me excited to be going back to work, although the minute Liz saw the rehearsal dates it got a little sticky. Just as had happened when Milo was born, I'd got myself a job that was going to take me away from home days after the twins were due.

'We'll manage,' Liz assured me. 'We'll have to.'

LIZ

I was thrilled for Kelvin that he got that part. I knew that it meant a lot to him to be on his way to the National Theatre. All his life, he'd dreamed of getting such a prestigious gig. There was no way he could turn down that chance, even if we were having twins!

Whenever life throws us a situation such as the one we were facing then – with Kelvin having to be in London for rehearsals while I looked after the farm, Marnie, Milo and newborn twins – I think of my mum and my grandmother. I've always had strong female role models, who've taught me that with determination and a bit of preparation, we can handle just about anything. I think of Mum, looking after me and my brothers while she did her degree and climbed the career ladder at the Crown Prosecution Service. And I think of Nana, who was left to raise her small children alone in Ireland while her husband – my grandfather – was in England looking for work. When he eventually sent word to her that he'd got himself a job and found somewhere for the family to live, Nana followed him, bringing her three kids to England on the ferry with no one to help her, only to discover that the 'family home' my grandfather had found for her was in fact an attic. She had to carry everything – children, their things, water – up and down a rickety ladder several times a day. She didn't complain. That kind of determination to just get on with it must be in my genes. But, ultimately, what else can you do?

*   *   *

Before the twins arrived, there was lots to prepare. Together with Gilly, we'd tried to time our lambing season so that it started the week before hers, hoping that she could give us a crash course before she turned her attention back to her own ewes.

We had five Cotswolds and eighteen mules in lamb. The mules were cross-breeds – Bluefaced Leicester and Swaledale. Hill sheep. They can withstand harsher conditions than the Cotswolds. The Leicester component gave them a big build and a bigger frame, which should make them easy lambers. It was me who bought those mules at market – under remote instruction from Kelvin, of course. He was in Bristol filming when they came up at auction, so I tagged along with Jack and Gilly. I still had morning sickness then and spent the whole time feeling like I could vomit at any moment, so naturally the mules we wanted came up last. It wasn't easy staying until the end of the auction, but I got a real sense of achievement from having got those sheep into the trailer and back to the farm on my own. Yorkshire shepherdess, eat your heart out.

Two weeks before the first lambs were due, the ewes were brought in to a sort of 'maternity ward' in the shed that Kelvin had prepared with as much care as any new dad painting the nursery. At this stage, I had to take a step back because being around sheep when they're lambing can be dangerous for pregnant women, due to a virus pregnant sheep sometimes carry. I still kept a close eye on their progress from a distance.

As my own due date approached, I felt especially connected to those ewes. One of the horses stabled on our farm was pregnant too. It seemed the whole world was waiting for babies to be born.

*   *   *

Gilly had warned us that once lambing season started, everything would go completely crazy.

'Whatever you do, do not plan to go away during lambing season,' she told us. Expecting hundreds of lambs on her own farm, Gilly wasn't planning to get much sleep. 'You need to be there round the clock. You never know what's going to happen.'

We took Gilly's advice on board, but right before our lambs were due, Kelvin and I were invited to the Olivier Awards at the Royal Albert Hall in London. The Oliviers are the biggest awards in British Theatre. It was an invitation not to be missed. Surely the sheep would be OK for just one night?

Knowing that it might be a while before we were able to have such a big night out again, with our twins on the way, we were really excited to go. I wore a beautiful flowing sea-green dress by Zeynep Kartal. Kelvin looked dapper in black tie. It was great fun to dress up. We couldn't have looked more different than we do on any given day on the farm.

Before we set off for London, Kelvin checked on the ewes in the shed. They seemed happy and relaxed. There was no sign that any of them were ready to lamb yet. All the same, we'd got Kelvin's brother Brayden to come over and stay at the farm as back-up, just in case. With Brayden installed in the farmhouse, Kelvin was reassured that we could safely leave the ewes in the shed overnight.

'It's just twenty-four hours. Nothing's going to happen. Let's make the most of our last night of freedom,' he said.

*   *   *

We had a great time at the awards ceremony, catching up with friends old and new. Our good friend Jason Manford, who'd hosted the evening in his usual hilarious style, invited us to join him at an after-party in the swanky hotel where we were all staying, and we happily said yes.

As we arrived at the party, Kelvin got out his phone to check on what was happening back home. He'd installed a webcam in the sheep shed so that he could keep an eye on the ewes while he was away from the farm. Usually, that camera showed the most boring footage imaginable. Thirty ewes, just hanging out in a barn. Watching that footage was the digital equivalent of counting sheep to fall asleep. This time, however …

'Oh no.'

I saw Kelvin's face drop.

'What's happened?' I asked.

'Have a look.'

He passed me his phone. While we'd been at the awards ceremony having a wonderful time, not one but *two* ewes had given birth. There were two sets of newborn twin lambs in the shed.

'This wasn't supposed to happen. We've got to get back.'

But how? It was after eleven o'clock. The last train to Macclesfield was long gone. While all around us glamorous guests drank champagne, Kelvin and I were anxiously glued to the screen of his phone as we tried to get a better idea of what was happening all those miles away.

'Do the lambs look OK?' we asked each other.

It was hard to tell. The sheep shed was so dark.

Kelvin rang his brother right away to tell him he had to get out there and check on the new arrivals, but the call wouldn't go through. Then Kelvin's phone ran out of battery. We quickly tuned into the camera app on my phone instead. Were the lambs suckling? Kelvin worried that one of them seemed to be standing apart on its own. Had it been rejected by its mother?

There was no way we could enjoy the after-party while such drama was unfolding on the farm. But what could we do? We couldn't get home and we couldn't get through to Brayden. Perhaps his phone was on the blink. We decided that it was too late to call Gilly. She'd been there for us on so many occasions and she'd almost certainly have stepped in to help then, but we didn't want to risk waking her up. Not when she had lambs of her own to worry about.

While we were panicking, Jason Manford called to say that he was on his way to meet us.

'All right, guys,' he said. 'Be there shortly.'

'Actually, Jason,' said Kelvin, 'I think we're going to turn in.'

We decided not to tell Jason why we were leaving the party early. He probably thought I'd had enough, being so pregnant and all. But we were actually in for a very long night.

Back in our hotel room, Kelvin finally managed to get through to Brayden, who said he'd go out to the shed at once. Kelvin and I could not take our eyes off the footage from the sheep cam. We didn't even get changed out of our evening clothes but sat on the edge of the bed in our finery, waiting anxiously for Brayden to appear on the screen. The moment Brayden appeared in the shed, Kelvin was ready with instructions.

'It's really simple,' he said. 'All you have to do is isolate the two ewes and their lambs in the pen for new arrivals …'

Simple for Kelvin, but of course Brayden had zero experience of wrangling sheep. He'd been to visit us on the farm before, but he'd never got up close and personal with the animals. He didn't know anything about the protocols Kelvin had worked out over months of trial and error. Sheep can be hard to corral at the best of the times, given their jumpy nature, but a ewe protecting her newborn lambs has even more reason than usual to be fearful.

Suddenly, the sheep-cam footage looked like a sequence from a Seventies comedy show, as Brayden chased the ewes around the shed, desperately trying and failing to persuade them to go in the direction he wanted, while Kelvin shouted increasingly exasperated instructions into the phone.

'Get both the lambs in the corner. The ewes will follow. Just concentrate on the ewes. Get ready to shut the gate the minute they're through …'

It took poor Brayden more than an hour to get the two ewes and their lambs into the smaller pen. He was exhausted. So were we from just watching the action on my phone. We owed him big time!

*　　*　　*

The following morning, we took the first possible train back to Macclesfield. We'd spent one of the most glamorous evenings of the year worrying about our sheep. We were so happy and relieved to find that the lambs and their mothers were doing well.

As Kelvin looked over the ewes still to lamb, I made a cup of tea in the farmhouse kitchen and asked myself: were we secretly more at home in our wellies than in black tie get-up? Had our frantic night shown us where our priorities really lay? I decided it had. We would never again risk leaving the farm overnight so close to lambing. We were definitely farmers now.

KELVIN

Thanks to Gilly, I had a protocol set up. I knew that when a sheep lambed, I had to spray them underneath with iodine to keep things germ-free and make sure that the new lamb sucked. I developed a simple system for numbering the lambs as they arrived. Gilly has her own system, so it was only right that I developed my own too. Single lambs got an odd number: one, three, five and so on. Twins were even: two, four, six. Triplets were noted down by letter: A, B, C ...

After successfully delivering a couple of lambs, I had my

first more challenging encounter. I found a ewe with her bag (the placenta) hanging out. She was panting and clearly struggling to give birth, which can be common with shellings (a first time lamber).

Not knowing what to do, I rang Gilly, and Jack came over to help. We caught the ewe and lay her down. She wasn't pushing.

Jack told me that I'd have to get properly involved. The lamb was in the right position but quite tightly in place. I tried pulling its feet, but it wouldn't come. I was full of adrenalin, but Jack was calm. I followed his instructions.

'You've got to put your fingers up its bum and push the head down.'

'Thanks, Jack. That sounds lovely.'

But I did as he told me and it worked.

Back when I was in *Emmerdale*, I had played the farmer, but I never would have expected that one day I'd be getting quite so involved in the business end of a sheep. It was not a moment I would have signed up for, but when it actually happened it was a brilliant experience. I'd safely delivered a lamb. As I watched it get to its feet, I was bursting with pride.

But, of course, no sooner was one lamb safely born than another one needed help. Two ewes had given birth while we weren't looking. One had twins and the other had a single. When I went into the pen, I saw one ewe in each corner with one lamb a piece and a lonely lamb in the middle, which was being neglected by both mothers. I had no idea which ewe was the poor lamb's mum.

I took a punt and put the lamb in an adopter pen along with the ewe that seemed most likely to be the mother. The adopter pen holds the ewe in place so that she can't go anywhere, while the lamb gets to suckle. After a couple of days, it worked; the lamb was successfully sucking on the ewe and the ewe had accepted her as her own. But talk about learning on the job.

*   *   *

For the most part, our first lambing season went pretty calmly and smoothly. We'd thought it would be chaos once it got going but there were days when nothing happened at all. Gilly was exhausted because she had so many more sheep.

We only lost two lambs. One of them was the smallest of triplets, born so small and frail that it had to be fed by hand. Marnie and Milo took turns to give the lamb a bottle and even named her. I put a heat lamp in her pen to keep her cosy. For several days we watched that lamb 24/7, willing it to grow strong enough to survive.

We would take it in turns to go out to the lambing shed. Late one night, Liz discovered that the tiny lamb had died.

Though I'd known it was a long shot that the lamb would survive, when Liz told me what had happened I got upset – more than I thought I would. I felt we'd let the little creature down.

It turned out that it had a condition called 'watery mouth' – a bacterial disease that affects newborn lambs. I was gutted when we found out. Gilly reassured me that there was nothing more I could have done, but I berated myself for ages for not having spotted that the lamb had an infection.

'You'll know for next time,' Gilly told me kindly, but the experience was hard won.

I still think about that little lamb, but ultimately we considered our first attempt at lambing a success. From twenty-three ewes, we had thirty-two lambs. And thankfully, they all arrived before our own twins.

CHAPTER THIRTY-ONE

WELCOME TO THE WORLD, BABY BOYS!

LIZ

During my first and second pregnancies, Kelvin was like a butler to me. I only had to glance in his direction and he'd be rushing to make me a cup of tea or give me a foot rub. My third pregnancy was very different. We were a busy farming family now. If I complained about any aches or pains in the hope of a shoulder massage or some sympathy, Kelvin would tell me, 'You'll be OK,' and carry on doing whatever it was the sheep, the pigs or the alpacas needed instead.

At the end of April, both Kelvin and I were very busy – with work, with the farm, with the children – but we planned that in the week before the twins were due to arrive, we would drop everything that didn't need to be done right then and just concentrate on getting ready for the big day. Because we were having twins and they would have to be delivered by caesarean, we knew long in advance what date they would be born. I was booked into Wythenshawe Hospital for the beginning of May. There were all sorts of things to prepare before then: there were two cots to be built; Kelvin was determined to clear out the boot room, which has been used as a dumping ground ever since we moved in; I wanted to get my roots and my nails done – it would probably be the last chance I had to pamper myself for a while!

But with just over seven days to go, I started to think that it was unlikely I'd be able to hold on until the day the hospital had planned for me. I was so big. It seemed impossible that I had another whole week to get through. I was starting to get Braxton-Hicks contractions.

On the Saturday before my due date, I woke at five in the morning to find that the bed was wet. Had my waters broken? I wasn't sure. With both Marnie and Milo, my waters hadn't broken, so I didn't know what to expect. Perhaps my bladder had finally given up on me after nine months of double the usual pressure.

Kelvin wasn't in the bed beside me. In the middle of the night, Milo had come into our room, wanting one of us to sit in his room with him until he fell asleep. Kelvin had done the honours and ended up getting into the empty bunk bed in Milo's room and falling asleep there, so he didn't disturb me by coming back into our room. When I went to tell him what was going on, he was still conked out. I gave him a shove.

'I think the twins are coming early,' I said.

Kelvin, only half-awake, said, 'Ah, right. You better ring the hospital.' Then he turned over and went back to sleep.

I rang the hospital, telling them that I wasn't sure whether my waters were breaking or whether I'd lost control of my bladder. They advised me to come straight in – which was all very well but it was five in the morning and, because we hadn't been expecting to go to the hospital that day, none of the support we had arranged for our official due date was in place. Instead, we'd been planning our last weekend as a family of four. We were going to get chores done before taking Marnie and Milo to see *Beauty and the Beast* on stage in Manchester. Even if I rang my parents, they wouldn't be able to get to us for a while and I was increasingly feeling like I needed to get to hospital quickly.

I packed an overnight bag and went back into Milo's room. Kelvin was still in the bunkbed.

'I'm going to drive myself to the hospital,' I told him. 'Get your parents to look after Marnie and Milo, then come and meet me there.'

Kelvin wasn't awake enough to argue with the plan. Neither was he awake enough to argue when I took his car: a brand-new people carrier – a seven-seater Land Rover

Discovery that could fit us all. I knew he would have protested if he'd thought about it.

I hauled myself into the driver's seat. All the time I was still leaking. I felt far from attractive. When I got to the hospital car park and got out of the car, there was no longer any doubt. My waters were breaking. They were breaking all over the driver's seat of Kelvin's new car!

In such a state, there was no way I had time to find the car park ticket machine. I'd just have to get a fine. I waddled as fast as I could into the hospital with one of Kelvin's hoodies wrapped around my waist to hide any wet patches on my leggings.

As soon as I got into the maternity unit, I was taken for a scan, which confirmed that the twins were ready to come out, even if I wasn't ready. There was no point waiting. I'd be taken in for my caesarean ASAP.

I called Kelvin and woke him up again. He told me that after I left, Milo had asked for a cuddle and they'd both fallen back to sleep.

'Well, try not to fall asleep again. This is it,' I told him. 'You need to get here. I'm serious. Now!'

KELVIN

Perhaps it was the fact that it was so early in the morning that meant I was slower to cotton on to what was happening than I would have been. When Liz told me she was driving herself to the hospital, I didn't think that meant she would be staying there. I told myself she would ring me to let me know what the doctors said about the leaking.

I had it in my head that we had another week. In many ways, we'd lost track of where we were in the pregnancy. On the one hand it seemed as though Liz had been pregnant forever. On the other, it suddenly seemed as if we'd just been told. So much had happened in the space of nine short months.

Hearing the tone of Liz's voice when she said, 'You need to get here … Now!' I snapped awake. She was her usual calm self, but I could tell she wasn't messing. Her final instruction was, 'Bring two bobbles for my hair.'

I got the kids up and gave them their breakfast. I packed myself an overnight bag. Ginger the dog was booked to stay with Gilly. Thankfully she was happy to take Ginger early. Marnie and Milo would go to my mum and dad. To save me having to drive all the way to Oldham, then back to Wythenshawe, Mum and Dad offered to meet me at a

Tesco service station on the motorway. I handed the kids over to them there. It was a surreal moment.

Everything was so hectic that it wasn't until I was in the car on my own, heading towards the hospital, that it hit me. I felt the excitement building inside me. I was on my way to meet my sons!

*   *   *

I got to the hospital to find that they were digging up the car park. I must have looked panicked because the work-men directed me straight to the staff car park. I ran into the maternity unit with my heart racing. What had I missed while I'd been in the car?

The moment I saw Liz, calm descended. She has a way of bringing peace to any amount of chaos and that's what she did now.

'Have a good lie-in, did you?' she teased me.

We had a laugh and a joke as we waited for the midwife to come and check Liz over.

'You should be in theatre at about midday,' was the midwife's view.

Midday came and went. In the end, it was gone two by the time we were taken to another ward and prepped for the adventure ahead. We got dressed in scrubs, just as we had done when Milo was born. The midwife told me I didn't need to wear the matching hat but I put it on anyway for comedy value when Liz and I took a last selfie of ourselves as parents of two.

How different this weekend was turning out to be from what we'd expected. When I met Mum and Dad at the service station to hand over Marnie and Milo, I also handed over the tickets for *Beauty and the Beast*. The children had been really looking forward to going so I told Mum and Dad to make sure they got to see the show. As we were going into the operating theatre, Marnie and Milo were going into a much more glamorous theatre in Manchester with my youngest brother Brayden and his girlfriend, Alexa.

After the calmness of the ward where we'd waited for Liz to be checked over, the operating theatre was hectic. Ten medical staff were setting up for the caesarean. As Liz was fitted with a cannula and given an epidural, I told the anaesthetist what we had expected to be doing that afternoon.

'*Beauty and the Beast*? I think we've got that soundtrack here.'

He had someone put it on.

LIZ

Though I'd been through a caesarean before, getting ready to deliver the twins in the same way was still an emotional moment. As I was poked and prodded and prepared for the operation, I couldn't help but feel a little overwhelmed. I trusted the medical staff absolutely, but there was still the anticipation of the unknown.

I'd been given the epidural but I could still feel my legs, so I began to think that maybe it wasn't properly taking effect – I mean, if it was, I wouldn't be able to feel them. I told the anaesthetist.

'Are you worried that the epidural isn't working?' he asked.

'Now I'm worried that *you're* worried it isn't working,' I said.

'It's working,' he assured me. 'It's normal for it to take a little while.'

The medical staff went on calmly setting up everything they needed. At some point, the anaesthetist tested the effectiveness of the epidural with a prick test. I didn't feel a thing.

Next, a big sheet was set up so that I wouldn't be able to see what was going on below my waist.

'Don't look down there,' I told Kelvin.

'What are you trying to hide?' he joked. 'I've seen it all before. There's no exclusivity. That's marriage.'

To be honest, I was more worried about Kelvin getting squeamish and wanting to pass out.

With everything finally in place, the medical team began the operation and at half past three precisely, the first twin was born.

As soon as Milo was born, he was placed straight on my chest so that we could bond immediately. With the twins, it was different. The operation was far from over. Twin One was put into one of the incubators that had been set up at the side of the room. Kelvin made sure he was with the first baby as he was put into the incubator. He whispered into his ear and kissed him, just as he had done when Marnie and Milo arrived.

'Daddy's here,' Kelvin told him. 'It's OK. Daddy's here.' The sound of Kelvin's voice seemed to calm him, while all around us the medical team carried on with the job of delivering Twin Two.

Twin Two, as he was called then, was delivered a minute later. He had to go on oxygen for a while. It was hard to be stuck on the bed, unable to get to my babies while the medical team were stitching me up, but within twenty minutes, both my boys were in my arms and calmness had descended again. I'd thought that having two babies to hold would be difficult, but it felt like the most natural thing in the world.

It was a wonderful moment. After the literal out-of-body experience of giving birth, you'd expect the first few hours to be crazy and hectic and have a soundtrack of non-stop crying. But it wasn't like that for us at all. It was really quiet. It felt as though everyone was settling down. Perhaps it was because we were all in shock – us and the babies.

We were taken back to the ward, where a nurse brought me and Kelvin tea and toast.

'This is going to be the best toast you've ever eaten,' she said.

It was pretty good.

We couldn't stop looking at our new babies. They were so perfect.

'We've done it,' Kelvin and I said to each other. 'We've done it. This is it. We're parents of four.'

*   *   *

We were in hospital for two nights. The morning we drove home was beautiful. That drive through the countryside felt like another precious gift of peace to be savoured. It felt as though the hills were welcoming us home and it was hard to imagine a time when we hadn't lived there.

Marnie and Milo had yet to meet their little brothers and we looked forward to the moment when we would all be together for the first time. As it happened, though, Gilly was the first person to greet the twins on their return home. She had been keeping an eye on the farm in our absence. Thankfully, the last of our ewes had given birth to her lambs the night before we went into hospital. Like me, she'd had twins, though hers were a boy and a girl. Something must have been in the air, because the day our twins were born one of the horses also gave birth. The farm was bursting with new life.

Gilly had come by the farm to drop Ginger off. We couldn't let her go without introducing her to the two newest Fletchers.

Then, minutes later, Kelvin's parents arrived with Marnie and Milo. The well-choreographed introduction we had hoped for was immediately out the window. Marnie and Milo charged in, full of noisy excitement. As Kelvin described it, it was like Christmas morning, when the kids see their presents too early and you can't stop them from diving in. Kelvin likes things to be done properly, but even he understood that there was no point trying to persuade the kids to take this moment slowly and quietly. Their excitement was precious. And the twins might as well get used to the chaos from the start.

Eventually, we persuaded Marnie and Milo to sit down so that each of them could have one twin on their lap. As they sat there marvelling at the new babies, my heart was full of pride.

'Look at our family,' I said to Kelvin.

It took a while for us to come up with names for the twins. Of course we'd talked about the possibilities before they were born but, as with Marnie and Milo, we wanted to wait until we'd met them to make the final decision. We wanted to carry on with the 'M' theme. We both like names that have proper meanings. I thought it would be nice to use an Irish name. On that frosty early-morning dash to Wythenshawe Hospital, I'd found myself driving past the Con Club, where my nana used to go to play bingo. I hadn't been past that club in years, and suddenly happening upon it made me feel like the spirit of my nana was with me and the twins.

We asked Marnie and Milo for their suggestions. Marnie said she liked 'Nico' and 'Connor'. We've got no idea how she came up with those names. Perhaps she'd heard them at school. Meanwhile, Milo suggested 'Toy Story' and 'Lion King', which shows what was going on in his head at the time.

After a couple of weeks, during which we tried out all sorts of combinations, we settled on Maximus and Matteuz. We've asked Jack and Gilly, who have become such good friends to us, to become godparents. It's been especially wonderful to see Jack, who is still a man of few words, bonding with our little boys.

We're really looking forward to the christening, which will take place in the village church. Mateusz has the middle name Kelvin. Meanwhile Maximus is going to be christened Maximus Crowther, after our dear old friend.

KELVIN

It's been five years since Crowther died but I still hear his voice in my head when I'm facing a big decision. I still think of him when something good happens and I want to share it – like the birth of our beautiful twins.

I was gutted that Crowther didn't get to see me in *Strictly*. Yeah, he would have taken the piss out of some of my outfits, but I know he would have loved the chance to be in the studio audience. He always used to tell me, 'You'd be great on *Strictly*.' He would find it funny that Liz and I have written this book too. Crowther and I used to talk about writing a book together one day, detailing all the adventures we'd had over the years, though we'd have to wait so we didn't embarrass our parents. We were going to call it *Never Drinking Again*. I can imagine him advising me on the title of this book. 'How about *Never Farming Again*?'

As I watch Marnie, Milo, Maximus and Mateusz growing up, I wish with all my heart that Crowther was around to see his own son pass the same milestones. We see Crowther's son as often as we can, and I try to make sure he hears what a great man his dad was. As he gets older, I'll tell him some of the wilder stories and I hope they'll make him laugh. I think he's already got a lot of his dad's personality. Sometimes, when he looks at me in a certain way, I can see Crowther's eyes smiling back at me in his son's young face.

I'll always miss Crowther. We all will. We were inseparable, just like twins. That's why we chose Crowther and Kelvin for the twins' middle names. Maximus and Mateusz growing up together will be like Kelvin and Crowther, part two. He lives on.

CHAPTER THIRTY-TWO

WE'RE YOUR PEOPLE

KELVIN

Having the twins has made us ask ourselves how we've evolved on our parenting journey since those early days with Marnie.

Thanks to the thirteen-year age gap between me and my youngest brother Brayden, I grew up with plenty of experience of babies and was always pretty good with kids – I love their eagerness and sense of mischief – but looking after your little brother is not the same as raising a child of your own. That's an entirely different level of responsibility.

In some ways I'm quite laidback as a parent. I'm the soft touch. Marnie and Milo know that if they ask Liz for a biscuit, the answer will probably be no. If they ask me, it's a different story ... Milo whispers, 'Don't tell Mummy.' I'll sometimes end up getting a telling-off from

Liz but seeing the delight in the children's eyes makes it worth it.

I'm not the kind of dad who gets stressed out when they're taking the kids round the supermarket and they keep picking things up. If they're not in danger of breaking something, what's the problem? You've got to let them explore. You've got to tell them that there are some things they can touch – a tin of beans, for example – but others, like fresh fruit, that they shouldn't. You have to tell them how the world works with calmness and clarity. You've got to let them be kids. They will always be the best friends I've ever had, but sometimes I have to be prepared to be temporarily disliked.

Liz and I have similar values but a different approach.

LIZ

For me, it's all about the emotion. Security is the most important thing. You can give children a sense of security by sticking to a simple routine. But it's a lot deeper than that. It creates a sense of familiarity that nurtures them and makes them feel safe. So bedtime needs to happen at the same time every night. That sort of thing.

Bedtime is one area where Kelvin and I do differ. I see it as a moment to wind down calmly. Sometimes Marnie,

Milo and I listen to a meditation tape. Then Kelvin comes in like Mr Tumble and suddenly everyone is wrestling on the floor.

'Kelvin!' I yell. 'Now is not the moment to start wrestling!' But the kids love it and I know that sense of fun is important too. It's what the children will remember in years to come.

As I've already mentioned, before we had kids I imagined that when we did we would raise them in a similar way to how Kelvin and I had been raised. Doesn't every parent want their children to have all the good things about their own upbringing? I wanted to live somewhere that had the same feeling as the Thorp Farm Estate, where they could play out with their friends at weekends and after school, knowing that all the adults would be looking out for them.

I'd thought that growing up on a farm would be isolating and perhaps even boring. I know now that I was wrong about that. Marnie knows so much more about life than I did at her age. Sure, Kelvin and I were streetwise, but Marnie and Milo have already experienced things that blow my mind. When we let that Texel tup out into the field and he started serving the females, I was mortified that Marnie was seeing what was going on.

'What's he doing, Mum?' she asked me.

But now I realise that I was wrong to be embarrassed. Marnie often talks to me about things she's seen around the farm – matters of life and death. She understands the cycle of life and the responsibilities that go along with farming. She often comes out with things that seem wise beyond her years.

'Mum,' she'll say, when we're talking about the people she meets at school, 'everybody's different.'

She is growing into a truly compassionate human being.

I hope we give all our children the confidence to be whoever they want to be. I hope that they always know they'll have our blessing whatever path through life they choose. Why shouldn't they try everything they want to? Marnie loves to muck around on motorbikes, but she also loves to dress up and dance along to clips of her dad's performances on *Strictly*. Milo has also picked up Kelvin's love of all things motorised, but he likes to cuddle up and sing along with me too.

Having my own children makes me treasure how Mum and Dad helped me to navigate the world. Trying to explain life to a three-year-old is really hard. You want to be able to give clear and honest answers to all their questions about life. However I was feeling, Mum always had a song – usually an Irish song – that fitted the moment. I wish I'd learned them all!

KELVIN

We know our children are privileged, in the sense that they have wonderful opportunities, but we also want them to have a work ethic. We hope that we'll be able to raise them to be level-headed and keep their feet on the ground. I love their sense of adventure. I love that they're into everything, whether it's mud, motorbikes or make-up.

We're not into material things. It's nice to dress up when we go to an event in London, but I'd never dress the children in head-to-toe designer clothes. We let them express themselves. If that means Marnie goes to school with wild hair sometimes, then that's OK.

Occasionally, at school drop-off, I find it hard not to be a bit embarrassed that ours are the scruffiest kids there. I was a pristine child. Because we didn't have much, we had to take care of what we did have and that meant not getting dirty or wet. I'm glad that my children don't have to worry too much for now and they're happy to go into a puddle or just run wild in the woods. It's a joy to watch them grow.

One recent afternoon, I picked Marnie up from school, and before she had even changed out of her uniform, she insisted that we went out to the woods in front of our house. On a short walk, we saw five deer. At one point, a butterfly fluttered by. At least I thought that's what it was.

'That's not a butterfly, Daddy,' Marnie told me. 'It's a moth.'

Her knowledge of the natural world and her eagerness to be in it reassures me that we made the right decision to leave our city life behind.

Like Liz, I've found that becoming a parent has only deepened my admiration for my own parents. It's only since becoming a father that I've understood just what they sacrificed for me and my siblings. Sometimes I'll just text them and acknowledge everything they gave me.

'Just wanted to say thank you,' I'll write.

My parents weren't educated. They couldn't help me do my maths homework or tell me how an oxbow lake was formed, but they gave me security. I could tell my mum anything, every secret. I always knew that, no matter what, my parents would never judge me. They let me be who I wanted to be, they supported me every step of the way, and for that I am truly grateful. I knew that if ever the world turned against me, they would be there, with open arms, ready to get me back up on my feet. Even if I never said it, I always felt that I am one lucky guy.

Another thing they gave us was self-confidence. Mum would cuddle us and say, 'You are absolutely gorgeous.'

Even if we didn't feel gorgeous at the time, her message of love sank in.

I want to give our children that same feeling of knowing that we'll always have their backs. As Liz often tells them, 'We're your people.'

EPILOGUE

JULY 2022

So here we are. It's our second summer on the farm and it's safe to say it's been a real adventure. We've gone from knowing nothing about farming to feeling like we've finally got the hang of things. We've made some mistakes, we've had plenty of mishaps, but we've learned to love our new life in the mud and the muck.

The arrival of the twins, and Kelvin having to travel back and forth to London to rehearse *Jack Absolute Flies Again*, have made the past few months the busiest and craziest we've ever had, but at last there's time to relax a little and spend a Sunday afternoon in the sunshine, picnicking in one of our fields.

It's the best feeling, to be sitting on a blanket with the twins, surrounded by buttercups, while Marnie and Milo race around us, enjoying the space and the freedom that only living on a farm can give.

'We're doing all right,' we say to each other. Our wild gamble has paid off. We're living the life of our dreams.

Marnie and Milo chase each other up and down, filling the air with their laughter. Ginger – a proper farm dog now – lazily snaps at a butterfly. We imagine how it will be in a year's time, when the twins are starting to walk. Is it time for us to get a horse? We're already thinking about getting some cows. We look out over our fields and imagine majestic Highland cattle grazing there. How about some more alpacas? But then we get a phone call … 'Your pigs are out,' says Gilly, and it's back to work again. It's all hands on deck – children and grown-ups – as we go back down the hill to chase our porkers back into their pen.

'Those flippin' pigs,' we sigh as we fold up the picnic blanket. 'Can't even give us an afternoon off.'

We know what Gilly would have to say about that. 'You don't become a farmer for an easy life.'

But it's a good life. Buying a farm is the maddest thing we've ever done, but we wouldn't swap our little piece of heaven in the Peak District and the happiness it brings us for the world.

OUTTAKES

KELVIN

The process of buying the farm took so long that there were moments when it felt like it might never happen. I'm a strong believer in the power of manifestation, so to help bring our farm dream into reality, Liz and I would sometimes visit the farm and the fields around it, just to soak up the atmosphere, connect with the surroundings and imagine what it would be like when it was ours. The farm was empty, and we were careful not to leave a trace.

Around Christmas 2020, when it looked as though we'd soon be going into another lockdown, it snowed. Not just a dusting, but proper snow. Enough for every slope in the area to be covered in kids on their sledges. Marnie and Milo were desperate to have a go, but there wasn't a sledge to be had. Liz suggested we just go out and slide down a hill on bin bags, but the nearest hill to us was rammed with teens and bigger kids. We didn't think it would be safe.

That's when we decided to pay another visit to the farm. The land we were buying included a hill that was perfect for sledging and, because it was private land, no one else would be there.

We bundled the kids into their snowsuits, which were so well padded they could hardly walk. They looked like a couple of Jelly Babies. We drove to the farm and headed up to the secret hill. It was brilliant. The kids loved zipping down the slope on their bin liners, though Liz wasn't too happy to see – as the snow was wiped away – that they were basically sliding on solid sheep poo.

Still, the kids had the best time, and we were really pleased with ourselves and our clandestine sledging trip. We thought we'd been really clever about it. We'd had a great afternoon and nobody knew we'd been trespassing on land that wasn't yet ours. We even walked down to the village pub afterwards. That was a cold walk. The kids were in snowsuits. I was in my trainers. I've learned a lot about dressing for the weather since then.

Since the farm has been ours, we've realised that we were naive to think nobody saw us that day. Our high field is visible for miles around and if anyone tries to walk through uninvited, we know about it. Liz is convinced we were featured on the local community's Facebook page. 'Watch out for cheeky sledgers.'

One of the first things Liz did when we moved in was to buy a couple of proper sledges. Of course, there's hardly been a flake of snow since then.

LIZ

When we first bought the pigs, Kelvin had big ideas for where they'd live on the farm. Grand designs, you could call them. He wanted to build a 'pig palace'. He mapped it out. He'd fence in a huge area, including part of the woodland, so that the pigs could roam pretty much wherever they pleased. It was a big undertaking to put up fences strong enough to keep them contained, though. As we found out as soon as we got them home, pigs are expert escapologists. You would be amazed at the destruction they can wreak with their snouts as they root about for food.

That last fact was concerning me as Kelvin put the finishing touches to the pig palace, which, as he'd planned, encompassed loads of the oak trees the farm's previous owner had planted. I'd seen the pigs go mad for acorns and roots, digging great big troughs with their noses that left the land looking like a JCB had passed through.

The BBC crew were at the farm at the time, filming Kelvin's hard work with the fences, so I took him to one side. 'Kelvin,' I said. 'I'm really worried. What if the pigs start lifting up the trees?'

Kelvin tried hard not to burst out laughing. 'Liz,' he said, 'they're pigs, not dinosaurs.'

Just in my eyeline, I could see the BBC crew cracking up.

Fortunately, Kelvin was right. Though the pigs are now six feet long and barely recognisable as the cute little piglets that arrived on the farm all those months ago, they're still happy in the pig palace and none of the oak trees has fallen down yet.

* * *

I didn't grow up around animals. Ginger, our dog, was pretty much the first animal I ever really got to know. When we first moved to the farm, although I knew we'd always planned to get livestock, I was still a bit nervous about what it would entail. But we needed to get the farm ready to pay for itself from the start, so Kelvin did lots of research into what we could do before we bought our first sheep.

An obvious answer was to rent out our paddock and stables. Kelvin put me in charge of that. I posted a Facebook ad on a local community board and was really surprised by how many people got back to me. We could have filled the stables ten times over. In the end, we rented them to a family with eight horses. They would come every day to feed the horses, make sure they were exercised and groomed. All Kelvin and I had to do was make sure the stables were maintained.

On the day the horses moved in, I couldn't quite believe it had been so easy.

'This is it,' I said to Kelvin. 'We're starting to make a living already.'

I felt quite proud that I'd managed to get the stable filled. All the same, it was strange to go to bed that first night knowing there were eight horses just across the courtyard. It was an odd feeling, having grown up on various housing estates, to be in the middle of nowhere with eight enormous, unpredictable animals as our nearest neighbours.

'Don't worry, Liz,' said Kelvin. 'You'll get used to it.'

It was a bit of a sleepless night. The wind howled around the farmhouse. It wasn't until the early hours that I was able to drop off. Then, as the dawn broke, I was woken by the strangest sound. An eerie clopping and snuffling, right outside. Still in my pyjamas, I pulled back the bedroom curtains and shrieked.

Eight great big horses had their noses pressed up against the window, staring in at Kelvin and me like we were the ones in a stable.

'Kelvin!' I yelled, though he was right behind me. 'What the hell are we supposed to do now?'

That was the start of a steep learning curve in dealing with animals. Still in our pyjamas, we ran out into the garden and began the long process of getting those horses back into the paddock. How could we get them to follow us? Kelvin tried to encourage them, as if they were school-children. I flapped my arms from the side. It was a bit like herding … well, it was exactly as you imagine herding horses to be when you're someone who's never been near one.

We worked out that the high winds must have blown open the gate from the paddock overnight. Another lesson learned – always check that the gates are secured. Since then, we haven't had another early-morning equine wake-up call. The sheep, the pigs and the kids, however? Well, that's another story.

KELVIN

From an early age, I've been mad about all things automotive. If I'd had access to a tractor when I was a little kid, I'd have been in heaven. So when Marnie asked if she could have a go at driving our first tractor, of course I was happy to say yes. I got into the driving seat and she climbed up onto my lap. With the speed of the tractor limited to just two miles an hour – barely walking pace – we set off with Marnie at the wheel. Just as when she first started herding the sheep, she got what she needed to do instinctively. I was in awe. With the soundtrack from *Matilda the Musical* playing from the speakers and a flat cap on her head, Marnie drove that tractor like she'd been born at the wheel. She spun the wheel with the heel of her hand, like she was a professional lorry driver. When she turned to look at me, she grinned as if to say, 'Look, Dad. This is what I do now.'

Marnie and Milo have really embraced farm life and that's in no small part due to our wonderful neighbour Gilly. The relationship Gilly has with the children is great to see. I think Marnie might have picked up some of her driving techniques from Gilly, who is always dashing around on her quad bike, filling the air with her familiar high-pitched sheep call as she goes. As soon as she comes into our farmyard, the kids are clamouring to go out on her bike with her. In fact, we have to admit to being just as pleased to see her as the kids are.

'Kids, Gilly's here!' we yell as soon as we hear her calling her sheep, and the kids race to pull on their wellies. Gilly gets a couple of mini sheep-hands on the back of her quad bike and we get a few minutes' peace.

*　　*　　*

Since we've had the twins, getting out of the house to go anywhere has become a real performance. A family outing requires at least an hour's prep. That's how I ended up taking Marnie and Milo to the local summer fête on my own, while Liz stayed at home with the little ones. I thought it would be a great way to kill a couple of hours. There were loads of planned activities for children at the fête, giving me the chance to sit down and have a drink and a chat with some of the local farmers.

We've been doing our best to fit in around here. Liz was recently invited to join the local ladies at a line-dancing class, which was unintentionally hilarious when the teacher, who took himself a bit seriously, banned the ladies from laughing as they danced. We all know what happens when you tell someone they can't laugh.

I was pleased that getting to know the men at the fête didn't involve having to learn a dance routine. I just sat down among them and hoped I blended in, wearing my brand-new Barbour, my wellies and a flat cap. Maybe I needed a bit more sheep shit on the wellies … but I thought I had the farming style nailed as I nodded along with the conversations about animals, weather and heavy machinery.

The kids were having a great time with a raucous game of Pass the Parcel. When that was finished, the fête organisers put on some music. My heart sank as I heard the first notes of 'Macarena' and one of the lovely ladies who'd been looking after the children beckoned me to come and join her.

Oh no, I thought. I'd been blending in with the old-time farmers, but this was about to blow my cover.

'Come on, Kelvin. Show us how it's done!'

It's the other 'Curse of *Strictly*'. You never can tell when you're going to be asked to strut your stuff. You're not safe even when you're just trying to have a quiet chat at a farm show. My new friends nudged each other, no doubt thinking, *This should be a laugh*.

But you can't take yourself too seriously when kids are involved, and Marnie and Milo were already on the dance floor.

'Daddy!' they called. 'Come and dance.'

In my Barbour and wellies, I wasn't exactly dressed for doing the Macarena, but I tossed my flat cap to the audience in a move that would have impressed even Craig Revel Horwood and threw myself into the dance. And you know what? Seconds later, the dance floor was heaving. Turns out everyone loves to dance. Even farmers. And as my old life and my new life collided in that marquee, I realised I had never been happier. I was the Dancing Farmer. Maybe there's a show idea in that.

ACKNOWLEDGEMENTS

Kids. We are pretty sure you don't notice how crazy our life currently is – you're happiest when we're simply together wrestling, cuddling or feeding the pigs. But in years to come, maybe when you read this book, you'll no doubt wonder what on earth we were playing at. Well, everything we do is for you. We love you more than you could ever imagine.

Our agent Olly. As well as our family and closest friends, there has always been one other constant beacon of support. We will forever be thankful for the faith and belief you have in our family. It's an honour to have you by our side on this journey. Thanks to Chris for all your help with telling our story.

Jack and Gilly. At first you were our tenants, then you became our mentors and inevitably our friends. Now, you feel like family. Without your patience, kindness and bloody great sense of humour, this new life wouldn't be the same. We're so happy you agreed to be our children's godparents.

KELVIN

My first thank-you goes to my Liz. Your patience, commitment and sheer strength allow our family to navigate life with resilience and determination. As a mum you are simply incredible. At times you may feel like it goes unnoticed, but we do see and cherish everything you do. Without you, our little family wouldn't be the same.

Crowther. This isn't quite the book we had in mind, but maybe that's a good thing! Some of my happiest moments were spent with you by my side. I think about you every day. Miss you, lad.

Thank you to Mary and David. Firstly, Mary. I've got nothing but admiration for the woman you are. I couldn't wish for a better mother-in-law. Liz's best qualities echo your teachings, which brings me to you, David. Although few, the not-so-good qualities all seem to come from you: impatience, stubbornness and the ability to be loud. But for some reason, I find myself loving those qualities in you like I do in Liz. You are a credit to Daniel and Michael, and I hope I've made you proud and taken care of your daughter as you would have hoped. There is one good quality she got from you, and that's her sense of humour!

LIZ

Thank you to Karen and Warren for always trying to better your family. Karen from counting pennies on the floor to making the brave decision to move home. Without that, I wouldn't have met my future husband that day in school, and I wouldn't have the family I have today. It's your impulse-buying nature that Kelvin has inherited, and it's certainly made our lives exciting.

And to Warren. You are Kelvin's rock; he has you to thank for his work ethic and kind nature.